"The 'science wars' have raged over the question of what we know and how we know it. In *The Wonderful Myth Called Science* Professor Fred Bauer gives the most convincing answer I have seen."

Michael J. Behe
Professor of Biological Sciences

* * *

"Dr. Bauer assaults the monolithic myth of Science and in doing so challenges us to toss out long-held but flawed assumptions – and to use the intellects that come packaged with our bodies.

Gerard E. Goggins.
Author of *Half-Wits*

* * *

"The most valuable function of Philosophy is to provide an omnipresent drum beat that can be heard throughout the forest, uniting it in a coherent and meaningful harmony. Dr. Bauer invites us on a scenic tour through that great forest."

Kevin J. Carlin
Professor of Physical Science

***

"Physics, chemistry and biology are 'Physical Sciences.' Economics is called a 'Social Science' since it deals with people rather than nature, molecules or cells. But should it be called science? If you think this is an interesting question, you'll find an excellent answer in The Wonderful Myth Called Science."

Ronald J. Shadbegian, PhD
Professor of Economics
UMass Dartmouth

The Wonderful Myth Called Science

Fred Bauer

SOLAS Press
Antioch
2008

Permissions Dept
SOLAS Press
P.O. Box 4066
Antioch CA 94531
USA

Printed in the United States of America

Library of Congress Cataloging-in -Publication Data

Bauer, Frederick, 1933-
 The wonderful myth called science / Frederick Bauer.
   p. cm.
 Includes index.
 ISBN 978-1-893426-97-9
 1. Science--Philosophy--History. 2. Science--History. I. Title.

Q174.8.B385 2008
501--dc22

                                              2008000579

For Nancy

To tell why would require another book.

Foreword

It has often been observed that, as a whole, the academic enterprise has grown, or as some would say degenerated, into a splintered array of disciplines. The disciplines are further divided into sub-specialties within each division. If each discipline is a tree, the practitioner perches on some sturdy branch, squawking in a sophisticated technical jargon that only those perched on the nearest branches can hope to hear or understand.

Most of us spend our careers in the treetops. There are good reasons for this. The new growth produces the fruit that sustains us. However, the fruit of our scholarship is abstruse and so tends to separate us from each other even more. Even mathematicians and physicists, who may share the same tree at times, find it difficult to reach across the gap. The most valuable function of Philosophy is to provide an omnipresent drum beat that can be heard throughout the forest, uniting it in a coherent and meaningful harmony.

Dr. Bauer invites us on a scenic tour through that great forest. He guides us among the tangled roots seeking for the soil from which each tree has sprung. That common ground is our own human experience, both as individuals from our earliest consciousness and as inheritors of an evolving culture. On this journey, we are encouraged to look with fresh eyes and see the entire forest anew.

The journey winds among the trees — Physics, Mathematics, Psychology, and even literary analysis are stops along the way. Such a journey is certainly motivated by a problem and should naturally conclude with an answer. Dr. Bauer does propose one that is most reasonable but the final destination is only the place where the tour ends. The journey itself is engagingly and craftily presented, accessible to any reader who would care to undertake the trip. Enjoy the ride.

Kevin J. Carlin. Assumption College September 2007

Contents

Preface

There has always been a need for metaphysics as it was defined in 1892 by William James. "Metaphysics," he wrote in the epilogue to *Psychology: Briefer Course*, "means only an unusually obstinate attempt to think clearly and consistently." Such thinking is nowadays more frequently called "critical thinking." But more than metaphysics or critical thinking is called for. What is now needed is critical choosing guided by thinking that is *super-disciplinary*.

Our colleges and universities have many competent critical thinkers; but they are segregated. Our college and university libraries are filled with books written by these critical thinkers; but those books are shelved in different sections. What is the effect of segregation? Simply this; that however practical it is, it reinforces and hardens in educated persons' minds a cultural myth. "What myth?" you ask —the myth that there are collective bodies of public knowledge which are independent of one another. The error that every educated person is bound to absorb is that there are public bodies of collective knowledge known as science, philosophy, and theology (religion).

The opening chapter aims to replace that myth with the true outlook. In truth there are only individuals' non-public bodies of personally acquired knowledge. One way to replace the myth is through the application of Ockham's Razor— a principle that Einstein includes in his view of 'the scientific method.'

Einstein once described science as 'a refinement of everyday thinking.' Chapter II describes the best way to approach everyday thinking. It describes how to view an individual's 'higher education' as a development of his or her 'lower,' everyday-thinking knowledge.

ix

The pivotal chapter is Chapter III. Its three segments form a bridge between the first two and final chapters. It will ease the reader into a realization that the very act of reading is the best way to discover what all 'higher' learning is. The first segment introduces the fact that later learning cannot begin until the learner has acquired the complex set of beliefs that is called 'everyday thinking'— that is 'common sense.' Using that fact, it explains why reading makes up the bulk of every expert's later knowledge. Segment two shows how Einstein's primary distinction between sensing and conceptual-thinking applies to all later refinements of everyday thinking. The last segment develops the view that every learner has her or his unique, unimaginably-complex worldview or mindset. The reader can use that insight to discern how their mindset influences the way they interpret what they hear or read.

The worldview of the author is outlined In Chapter IV. Special attention is given to two mythical concepts: First, that science is a special body of truths intelligible only to experts. Second, that 'scientific' conclusions must be taken 'on faith' by non-experts who have not mastered the esoteric mathematics. To help readers 'see through' both these errors, the chapter tells why two great scientists, Einstein and Bohr, fundamentally disagreed with each other. It then goes on to explain what mathematics really is. The book ends by applying the worldview outlined to the act of reading. It focuses on what "I see colors" means.

I ask the reader to bear with the extensive use of the first person singular. That is part of the argument. My learning and choosing has all been individual and personal, and whoever you are, your reading, interpreting, and choosing must be the same.

It is self-evident that a work of this kind depends substantially on the help of persons too numerous to mention. However, special mention should be made of my many inquiring and questioning students who over the years have motivated my inquiry. Also many of my faculty colleagues have extended help and valuable criticisms for which I am truly grateful. The remaining errors are entirely mine.

F. Bauer, Worchester   July 2007

The Wonderful Myth
Called Science

CHAPTER I.

Science Is a Myth

The aim of science is, on the one hand, a comprehension, as complete as possible, of the connection between the sense experiences in their totality, and, on the other hand, the accomplishment of this aim by the use of a minimum of primary concepts and relations.

A. Einstein, *Ideas and Opinions, p.286*

A. THE PRESENT

A fact.

Science doesn't exist. It's that simple.

Of course, you'll never guess that, if all you go by are today's newspapers and magazines. They refer to science as routinely and nonchalantly as they refer to the economy, the government, and the weather, and this is a big part of the problem. Those things — the economy, the government, and the weather — don't exist either.

In fact, practically none of the things the newspapers and magazines refer to actually exist. We have ideas of science, the economy, and so on. They, the ideas, are the only reason we can read and understand what newspapers and magazines tell us.

But having an idea of something isn't enough to make that something real. That's just plain, everyday common sense. After all, we have ideas of Santa Claus and elves, but that doesn't make them real. We can enjoy *Star War* movies and *Star Trek* TV shows, and we can even study the

Science Is a Myth

book entitled *The Physics of Star Trek*. But none of that makes the stories true. Darth Vader and Captain Picard are as mythical as Santa Claus, and as mythical as science.

A very slow conversion

Science doesn't exist. But it really isn't that simple. If it were, there' would be no reason to write this book. After all, the announcement has been made. It's true. It's so obviously true that, once it's been publicized, everyone can tuck it away in their memories and get on with their lives. That's what we do when we are told on the 11pm news that the Indians won 5-4 in the bottom of the ninth.

But it isn't that simple. Not even hearing the 11pm news and filing away the score is that simple. Earlier in the day, at least eighteen athletes had to spend well over an hour going through complex motions that might take at least 400 pages to describe in enough detail that someone who had never heard of baseball would understand what "The Indians won 5-4 in the bottom of the ninth" means. Prior to playing the game, each of those athletes had to be born, grow up, practice for endless hours to develop the exceptional skills that earned them a place on a professional, major league, baseball team. There isn't even anything simple about major league baseball, either. In fact, nothing in life is really 'just that simple.'

The non-existence of science certainly wasn't simple for me. I have no idea when the concept of science began coming into being for me. But by the time I had finished college and was packing to go off to graduate school for an advanced degree in philosophy, my concept of science was, I thought, quite well-developed. The most important thing about science — we usually called it "empirical science" — was its inferior status vis-à-vis philosophy and the queen of all disciplines, sacred theology. So, when I was advised by one of my former philosophy professors to concentrate on learning as much science as possible, because of its growing importance in the world, I took that advice.

I soon began reading the history of astronomy, relativity, quantum theory, chemistry, biology, mathematics, psychology, linguistics, etc. My experience was just the opposite of Thomas Kuhn's. As he tells in the 1962 preface to *The Structure of Scientific Revolutions*, he lost his faith in science as a slowly accumulating body of proven truth about nature. I slowly converted to that faith. It undermined my belief that philosophy and theology are solid bodies of truth impervious to supposedly transient theories developed in empirical science.

Not until many years later did I finish unifying my worldview. By then I had concluded that science really doesn't exist. But, of course, neither do philosophy or theology. The story of this conversion from compartmentalization to integration is a long and complicated one. This book will not tell that story. It will report only on my conclusions. The first of which has already been announced.

Science doesn't exist. In a funny sort of way, *it is just that simple.*

Why do we believe there is a thing called "science"?

If science doesn't exist, why do we come to believe that it does? To find the true answer, it helps to rule out the obviously wrong ones. For instance, it is not because each of us was first to make its discovery. I know I didn't invent science. Have you ever wanted to tell the world that you did?

If no one alive today invented science, why have we all come to believe in it? The answer, believe it or not, is easy. It's "The word." The word itself. *Science.* We grew up surrounded by people who believe in science. At some point in our learning career, they used the word in our presence. We heard the word, and we did what all newcomers to this world do. We developed an idea to go along with the word. Eventually we got in the habit of using the word ourselves. It didn't matter whether or not our idea was just like the idea of other people who used it. As long as no one challenged what we said, or as long as we did not hear people use "science" to say things which contradicted what we believed, it did not occur to us to wonder just what "science" did or should mean.

We form many, probably most, of our ideas the same way: from hearing words and names. Think of how we got our idea of Santa Claus.

3

Science Is a Myth

The people from whom we learned the names for twinkling stars, passing clouds, moms, dads, puppies and kittens, began at some point to use the name "Santa Claus" in our presence. From the things they said, we naturally — and without a second or critical thought — developed the belief that "Santa Claus" was the name of a man who runs a toy shop at the North Pole located at the top of the world, a man that knows who's been good or bad, naughty or nice, who delivers the toys found under the Christmas tree, etc. The only difference is that the people from whom we learned about science and Santa Claus knew Santa didn't exist, but assumed that science did.

Further evidence for saying that hearing the word "science" was the reason we grew up believing in science comes from noticing that no one has ever seen it. We learn that "star" goes with the tiny white pinpoints we see in the velvety black, nighttime sky. We learn that "cloud" goes with the fleecy white things we see drifting overhead during the day. We learn which of the things we see are called "mom," "dad," "doggie," and "kitty." But no one ever pointed to something and said "Look, there's some science."

Nor did anyone ever point to something and tell us it was the economy, the government, or the weather.

Why do we keep on believing science exists?

Neither Santa Claus nor science exists. But there is a difference. All of us learn that Santa Claus is a mythical person, that is, a person who doesn't exist. But nearly everyone who begins to believe in science continues believing in it. Why is that?

The question should intrigue us. Why do we not keep on believing in Santa Claus, even though it seems there is a lot of evidence for his existence? Pictures of Santa are everywhere after Thanksgiving Day. There are songs and even a famous poem about him. Around Christmastime, he visits malls and takes toy-requests from little children. He comes to Christmas parties and hands out presents. But still, we quit believing in Santa Claus. Why?

And why do we not quit believing in science, even though there is no visible or tangible evidence for it. Even people with PhD's continue believing in science long after they have given up their belief in Santa Claus. Why is that?

The answer is simple. It goes right along with the reason we start doing it. It's because everyone, it seems, believes in science. We are surrounded by people who believe in science, who talk about science, and who take the question, "What is science?" so seriously that they argue about its true nature endlessly.

B. THE PAST

A view from history

The idea of science was invented, not discovered. It is safe to say that twenty-thousand years ago, none of our human ancestors had the word "science" in their language. It was not only because the English language did not yet exist. They didn't use any equivalent word because they didn't have the idea.

What they most probably did have was the idea that some people in their village or tribe were smarter than most. That's just a guess, of course, but a very natural, common-sense guess. When we go to school, we quickly discover that not everyone is equally smart. Some classmates consistently get better grades. And, what sometimes makes us conclude that life is not fair, some of those with higher grades don't even study as hard as we do.

But, whatever we guess about our most distant ancestors who left behind no written records of what they believed, our more recent ancestors did leave records. Those records show that some people did eventually create a special idea for smartness. Our English word "science" evolved from the Latin word *scientia*. It, in turn, evolved from the Greek word "episteme." It was Greeks who converted the loose idea of individual people's smartness into the abstract idea of science.

Science Is a Myth

The best way to think about the invention of the myth called science is to picture yourself living in ancient Athens during the lifetime of Socrates. He died roughly twenty-four centuries ago, in 399 B.C. The intellectual life in Athens at that time was similar in many respects to the intellectual life, say, of Boston, Massachusetts. Just as in Boston today there are brilliant thinkers who dialogue with one another, so it was in ancient Athens. If we believe the books that can be found in any well-stocked library, Socrates had a passion for learning. He dialogued with many of the leading figures of the day. As his own thinking matured, Socrates began to believe that many of those leading figures were more clever than smart. Clever people can talk a lot, but they often do not know what they are talking about. Who, Socrates began asking, really knows what they are talking about?

There are many ways of asking that question. Which of the smartest people has more than just an opinion on this or that topic? Whose opinion is solid, because he or she has plenty of evidence to back it up? Whose opinion is so solid that it is no longer just an opinion, but knowledge? Whose view can we be absolutely certain is right? Who is the most intelligent person of all? Who is the wisest?

But wait. Are those all the same question?

The beginning of rigid definitions.

Are all of those questions really one, same question? They certainly don't look like it. Look at them. Notice the different words they use: smart, just an opinion, solid opinion, evidence, knowledge, view, absolutely certain, right, intelligent, and wisest. If you have some extra time on your hands, you can get out your dictionary and do a bit of sleuthing. See how each of those terms is defined. Check on whether the definitions overlap enough to be declared synonymous.

Of course, to be sure of your answer to "Are those different questions really synonymous," you will have to be sure about the meaning of "synonymous." Just for fun, I've spent the last five minutes checking four different dictionaries to see how "synonymous" is defined. All four begin their answer to "How is 'synonymous' defined?" by putting a limit on the words that can be synonymous: they must be words from the same language. But one tells me that synonyms are words that have *nearly* the same meaning. Another says they have the same *general* sense. Are "nearly" and "generally" synonymous? What about "meaning" and "sense"? Is the sense of "meaning" the same as the meaning of "sense"? Most of all, what does "same" mean? One human can be essentially the same as a second human, and they can save money by living in the same apartment. Clearly the first "same" doesn't mean the same thing as the second one. The first is synonymous with "similar," the second means the same thing as "not two, but one."

Let the preceding paragraph serve as an introduction to one of the most devilish of all questions. How can we be certain that we know what a particular word means? We need words — at least what appear to be words — in order to communicate with one another. But words are tricky, because they can be used to mean radically different things. "We can kill a bat with a bat" is an easy example. "Science" is no exception to the rule.

That is, not only is there no such thing as science. There is no single meaning to the term. "Science" can be the name for many different ideas related to thinking and speaking. One of my dictionaries offers two definitions, another gives five of them. And the people who believe science really exists disagree so thoroughly with one another about what it really is, hence what the word really means, that they write books to defend their own view against those of others. This is one of the complications that must be faced by anyone who sets out to prove that science as such does or doesn't exist.

As far as we can tell from historical records, the complication began with the same Athenians from whom we inherited the word. Like ourselves, the Greeks used a variety of closely related terms. Where we use sound opinion, knowledge, science, certain truth, and so on, they used *episteme, techne, phronesis, nous, sophia*, and so on. Socrates got in the habit of demanding exact definitions of terms. His protégé, Plato (427-347 B.C.), constructed much of his worldview on the premise that finding true definitions is an essential part of discovering the truth about

7

reality. Plato's most famous pupil, Aristotle (385 - 322 B.C.), was a great systematizer, and in the sixth chapter of his extremely influential book on ethics, he set out to give precise definitions for each of those five Greek terms.

Those three thinkers acquired a new habit. It was an outgrowth of something very natural, even instinctive. We are all capable of switching from a very loose use of vocabulary to being very precise with our terminology. Most of the time, we use terms very loosely. If we tried to think out the precise meaning of each word we use in normal conversation, it would take hours for even the most trivial exchanges. But we don't. We use "word," "name," and "term" interchangeably, just the same way we all use "same" for two utterly unsame meanings, just the same way we use "metaphor," "analogy," "simile," and "comparison" quite loosely unless we are writing a paper or exam for an 'Introduction to Literature' course.

But those who develop a passion for truth comparable to that of Socrates get in the habit of insisting on precision. As was noted above, it is a natural outgrowth of our tacit awareness of the difference between using words loosely and striving for precision. In everyday arguments, we find ourselves asking "Are you just saying that, or do you know it for a fact?" Or insisting "You don't know, you only think you know." The myth of science originated from this awareness of the difference between first-thought, loose use of 'knowledge terms' and second-thought striving for precision.

The same conversion from our growing-up, loose use of terms to classroom efforts to use words with self-conscious exactitude is at the root of today's major complication, the three-fold, mythical classification of truth or alleged truth. At first, we all use "science," "philosophy," and "theology" (or "religion") loosely. By the time we graduate from college, our vague ideas of the three will be separated by almost impregnable walls. But, like all of our contemporaries who have graduated from college, we each develop our own version of the classification. Our versions differ radically, except in one prominent way. All of them are mythical.

The Wonderful Myth Called Science

A warning

This brief history is being deliberately simplified. Its only purpose is to provide a background context for the thesis that there is no single, real counterpart for the generic idea signified by "science." Both Plato and Aristotle invented terms to distinguish the knowledge of people whose smartness extended to most topics from that of people who were smart only about one or two. They also taught that some specialized types of knowledge are more important or of greater dignity than other specialized knowledges. Those further distinctions are distracting details. Anyone curious about them will find enough resources in a well-stocked library to keep them supplied for years of study. The last thing needed here are more confusing details.

Not one, but two sciences: theology and philosophy.

The preceding introductory history suffices, then, for understanding what came next. Had everyone who lived after Socrates, Plato, and Aristotle stayed within the framework established by them, it is conceivable that we today would not gradually acquire the habit, quite early, of believing in science, as well as in something called "philosophy" and something else called "theology" (or "religion"). Everyone might have grown up thinking there was only one generic kind of smartness, and the only important question might have been "Which smart people know the truth?"

History did not work out that way, however. Four centuries after the death of Socrates, Jesus Christ was born. Like Socrates, he was a powerfully influential teacher. Like Socrates, he wrote no books. But, again like Socrates, he was soon immortalized by books about him and what he taught.

Initially, his followers felt no need for any other wisdom than what Jesus had taught. In time, however, disputes arose about some of those teachings, especially two of them. First was the idea that there are three divine individuals called "the Father," "the Son," and "the Holy Spirit," respectively. It was not immediately clear how that could be squared with the traditional idea that there is only one god. Second was the idea

9

that Jesus was actually the eternal, divine Son who, at the some point in time, became genuinely human as well. Some Christians turned to Plato's theories for help in resolving the apparent contradictions. Others turned to Aristotle.

To make a very long and tortured story short, it eventually became customary to divide all expert theories into two groups. Experts in the one branch of teachings built their theories around what was written in the set of books now called "the Bible." Inasmuch as what was written in those books was about God, that group of expert theories was called "theology," meaning "scientific knowledge about God." Theories similar to those of Plato, Aristotle, and others, i.e., theories based not on divine revelations but on a direct study of nature itself, were labeled "philosophy," a word previously invented by the Greeks to mean "the love of genuine wisdom."

What is noteworthy for our purposes is this. Among the terms used by the medieval, Latin thinkers to define "theology" and "philosophy" was the Latin term *scientia*. Both theology and philosophy were defined as *scientia*, meaning something like "proven and therefore certainly true knowledge." The people with authority during the centuries leading up to Copernicus and Galileo maintained a close watch over those who taught anything that contradicted the official theology. So long as teachers of philosophy steered clear of theological ideas, they were free to think what they wished. That distinction between jealously protected theology and less restricted philosophy was threatened, once Copernicus, Galileo, and other great discoverers arrived on the scene.

A third science, now regarded as the only one.

The myth that there is a new kind of science or certainly true knowledge distinct from the two older sciences, theology and philosophy, resulted from the great discoveries made by individuals whose names most of us become familiar with. First among the best known is that of Copernicus, who was born in 1473 and died in 1543. After him, there were Galileo (1564-1642), Newton (1642-1726), Darwin (1802-82), and Einstein (1879-1955). Their expert knowledge has become the only type most people are now in the habit of calling "science."

The reason for no longer thinking of theology and philosophy as science is simple. Most people do not regard them as proven and therefore certainly true knowledge. Only the modern theories about the earth going around the sun rather than vice-versa, about Galileo's gravity causing bodies to fall rather than Aristotle's essence, about Newton's single set of laws to explain both the heavens and the earth rather than Aristotle's two separate sets of laws, etc., are viewed now as proven, hence reliably certain.

Like it or not, we must recognize that "by their fruits you will know them" is pure common sense. That is why, if anyone doubts the superiority of modern science over the two older so-called sciences, theology and philosophy, we have only to direct their attention to the 'fruits' of modern science, namely, the miracles of technology. Planes get us from one city to another faster than any horse can. Xerox machines reproduce ten full pages of writing faster than ten short sentences can be copied by hand. Satellites convey messages from one side of the globe to the other faster than a note can be delivered from one end of a street to the other. And, somewhat frightening to think about, thousands of people can now be killed with a single bomb in a fraction of the time it would take to kill them individually with a sword or a spear.

Facts such as those are all most people need to feel confident that only today's science is genuine science. Facts such as those are more than sufficient to explain why many people believe that Darwin's and not Moses' story of human origins should be taught in public schools. Facts such as those explain most people's puzzlement when they are asked why graduating physicists, chemists, and biologists still, at the present time, receive doctoral degrees in philosophy rather than in science. No one can dissolve that puzzlement who does not know something of the past.

However, people in the future will never discover the full truth about themselves and this world we live in unless they first clear out of their minds the myth called science.

And the additional myths called theology (or religion) and philosophy.

C. THE FUTURE

Ockham's Razor: replacing myths with true thoughts

Albert Einstein was named "Person of the Century" by *Time* magazine in its December 31, 1999, issue. The reasoning behind the designation was explained in an introductory essay which cast an eye over the twentieth century, picked out three major themes, singled out the third as the most significant, and finally pointed to Einstein as the person whose memory will forever be linked to that third theme. It read, "In a century that will be remembered foremost for its science and technology . . . one person stands out as both the greatest mind and paramount icon of our age."

Someday, however, Einstein may be viewed as the scientist whose writings helped to replace the myth of science with a clearer vision of the truth. His reflections on the relation of humans' minds to existing realities should eventually be seen as having as much value as his famous formulas for predicting sensations.

One of his most cherished principles was his belief that the simplest of comprehensive theories is the one most likely to be true. In a 1936 essay, "Physics and Reality," he expressed that two-sided principle as follows.

> The aim of science is, on the one hand, a comprehension, as complete as possible, of the connection between the sense experiences in their totality, and, on the other hand, the accomplishment of this aim by the use of a minimum of primary concepts and relations. (A. Einstein, *Ideas and Opinions,* p.286)

". . . a minimum of primary concepts and relations." The remainder of this chapter will show how, by adopting Einstein's two-sided principle of adequacy-plus-simplicity, anyone presently burdened by the myth of science can remove it in good conscience. They not only can but should remove it. Why? Because believing that such a thing as science in the singular exists violates the rule about a minimum of primary concepts. *It violates the principle of simplicity.*

Believing in as few things as necessary

Why believe in something extra named "science" in the singular when we already believe in millions of things called "people and their individual beliefs"? Why believe in a *Time* magazine in the singular when we already believe in millions of magazines in the plural, each with the word "Time" on its front cover? If it is possible to explain every sensation any of us ever experiences without believing in anything extra besides, over and above, or in addition to individual people, their individual beliefs, and their individual actions, why do it? It does nothing but create needless clutter in our thinking.

For instance, review the earlier paragraph which contained the following sentence, "Albert Einstein was named 'Person of the Century' by *Time* magazine in its December 31, 1999, issue." It is clear that, if any naming was done, it was not done by some invisible magazine in the singular. *Time* didn't choose to name Einstein the "Person of the Century." It was, we say, flesh and blood human beings who selected his name from the dozens of other names they were familiar with.

If, therefore, we begin taking what we hear and read about science and asking "What real things are being talked or written about?" it becomes apparent that, with few if any exceptions, those real things are human beings and their opinions, things we already know about and believe in. All disputes, even 'scientific' ones are like that. A dispute is a disagreement between individual people like bakers and butchers and candle-stick makers who think their own opinions — unlike the opinions of those with whom they disagree — are solid, backed up with sufficient evidence, and therefore true. Scientists' disagreements with other scientists are no different. There is no reason to believe in any 'more' in their case. That is, there is no reason to believe in a special thing named "science" in the singular — or in theology or philosophy.

Science Is a Myth

Analyzing statements about science is like analyzing poetry. "Science now does its thinking in public" is a statement found on page 66 of a May 22, 2005, magazine. It is obviously the product of someone's creative imagination. Thinking is done, not by science, but by humans. Thinking is not public. Only the spoken or written words used to express some humans' private thoughts can possibly be public.

This principle of not believing in more things than are needed to explain the facts needing to be explained has many names. The most popular one is "Ockham's Razor." It is named in honor of William of Ockham (1285-1347) who explained that human knowledge can be explained without invoking the existence of anything but individual entities. Because applying the principle often requires eliminating false beliefs and therefore reducing the number of things we believe in, it is also called "reductionism." Other times it is referred to as "the principle of parsimony," "the principle of simplicity," or simply "kiss" (for "keep it simple, silly").

This 'reductionist' principle is common sense, pure and simple.

As we will see in later chapters, the modern discoveries by Copernicus, Kepler, Galileo, Newton, Dalton, Mendeleev, Einstein, and Rutherford compel us to radically revise many of our most common-sensical convictions. But the revision itself cannot be justified except by other common-sensical convictions, even if we do not notice at first that we are using them.

The principle of simplicity, Ockham's Razor, and/or reductionism is one of them. We do not realize this while we are growing up. We learn to speak prose long before we ever hear the word "prose." We think about galaxies and brains and atoms, without noticing that our thinking follows rules of logic or psychology. But speaking prose, even if we never notice it, and thinking in ways described technically by logicians and psychologists, are obvious features of everyday, common-sense speaking and thinking. A few examples will make it clear that the principle of parsimony is also an ordinary, everyday feature of our thinking.

The Wonderful Myth Called Science

Consider, once again, our youthful belief in Santa Claus. We start believing that there is a jolly old man running a toy shop at the North Pole without ever going there to see for ourselves if Santa exists. We quit believing Santa lives on top of the world without ever going there to prove to ourselves he doesn't exist. If we review that whole 'Santa' episode in our early learning career, we may decide that the focus of that belief was those Christmas presents that magically appeared under the Christmas tree on Christmas morning. Where did they come from? Who put them there? Santa Claus did, we were told.

As we get older, however, we begin to notice things that perhaps we didn't pay any attention to previously. Santa doesn't always look the same when we see him at different locations. Entire aisles in various stores have shelves filled with the same kind of toys that Santa's elves make. When we go snooping in our parents' bedroom closet just before Christmas, we find the very things which later appear under the Christmas tree. Finally, Susie down the street or Johnny next door tells us that there is no Santa Claus. Once we can no longer doubt that mom and dad bought the gifts and put them under the Christmas tree, there is no longer any reason to continue believing Santa put them there. It makes no sense to believe that there are two different causes for one set of presents.

Similarly, there is nothing that people call science which cannot be explained in terms of real, individual people and their real, individual opinions.

Believers reluctant to surrender their long-standing faith that science exists will react to the 'Santa' illustration with a defensive "That's different." One of our most important psychological abilities is being able to see when things are different. Without it, we'd be unable to tell mom apart from dad or anyone else. Learning to believe in Santa is admittedly not in all respects the same as learning to believe in science. We do not first know that our parents put gifts under the tree and later add a superfluous belief in Santa. In the case of science, however, we do know about lots of people and we do know they have lots of knowledge, and we do know those things before we begin to believe in science. And so, it might be argued, science is different. True. But . . .

But they are similar in the way we start to believe in them. In each case, it seems there's more than what we already know. Were it not for the fact that we grow up surrounded by people who themselves believe there's something named "science," attend schools where things are learned that go far beyond everyday street knowledge, where the textbooks have additional names, e.g., physics, chemistry, biology, social science, etc., in their titles, we might not be so ready to believe there is something over and beyond individual people and their personal beliefs.

But that's not all. Stopping to believe in Santa is not so different from giving up old beliefs in exchange for new 'scientific' discoveries. Galileo learned Ptolemy's view of the heavens before he began to argue that it ought to be replaced by Copernicus's astronomy. The earth and the sun are 93,000,000 miles apart —a distance regarded as enormous in the 1600's. They cannot both be in the center of one and the same universe. Thus, the same common-sense intuition that tells us both Santa and our parents did not put the same set of gifts under the tree also led Galileo to conclude that, since Ptolemy's and Copernicus's astronomies could not both fit the same motions of the same bodies, it made sense to choose Copernicus's because it was simpler and explained more things. (Debates continue to rage about both of Galileo's reasons.)

Similarly, once Newton's contemporaries saw that his three laws of motion could unify our view of both celestial and terrestrial phenomena, they discarded Aristotle's incompatible pair of principles. If we accept Darwin's date for human origins, we simplify our thinking by discarding the story told by Moses, much the way earlier Christians adopted Ptolemy's view and discarded the biblical description of the earth as flat.

Of course, if someone is determined to believe in leprechauns, they are entirely free to do so.

The Wonderful Myth Called Science

Conclusion.

We must simplify our thinking. We do this by taking all of our opinions about science, philosophy, and theology (or religion), and grouping them under a single category labeled "theories" or "opinions." At times, it may be convenient to sub-divide our theories and opinions vis-à-vis the things they are about. But, regardless of what they are about, theories and opinions are theories and opinions.

The resulting simplification in our search for truth is immense. Instead of worrying about whether this or that opinion is scientific, theological, or philosophical, we need to worry in the future only about whether it is true.

CHAPTER II.

The Foundation: Everyday-Thinking Habits

The whole of science is nothing more than a refinement of everyday thinking. It is for this reason that the critical thinking of the physicist cannot possibly be restricted to the examination of the concepts of his own specific field. He cannot proceed without considering critically a much more difficult problem, the problem of analyzing the nature of everyday thinking.

A. Einstein, *Opinions and Ideas*, p.283

A. EVERYDAY THINKING 'IN GENERAL'

Einstein, like Huxley, was right about common sense.

One of most valuable lectures ever delivered in relation to 'science' was one given by Thomas Huxley in the second half of the nineteenth century. Huxley (1825-95) is most famous for his brilliant defense of Darwin's theory of evolution. His lecture, "The Method of Scientific Investigation," was addressed to an audience of British workingmen and opened with a bold thesis: "The method of scientific investigation is nothing but the expression of the necessary mode of working of the human mind." He then continued:

> It is simply the mode at which all phenomena are reasoned about, rendered precise and exact. There is no more difference, but there is just the same kind of difference, between the mental operations of a man of science and those of an ordinary person, as there is between the operations and methods of a baker or of a butcher weighing out his goods in common

scales, and the operations of a chemist in performing a difficult and complex analysis by means of his balance and finely graduated scales. (In S. Rapport & H. Wright, *Science: Method and Meaning*, p.2)

Huxley's insight must become our own. Before continuing to believe that there are different kinds of specialized knowledge acquired by different kinds of specialized methods, we must all take a closer look at the fantastically complex thinking we do so effortlessly and unself-consciously in our everyday lives.

If we need any encouragement, some advice from Einstein should supply it. He came to a conclusion even broader than Huxley's. And he didn't address it only to lay folk. He directed it to the experts as well.

The whole of science is nothing more than a refinement of everyday thinking. It is for this reason that the critical thinking of the physicist cannot possibly be restricted to the examination of the concepts of his own specific field. He cannot proceed without considering critically a much more difficult problem, the problem of analyzing the nature of everyday thinking. (A. Einstein, *Opinions and Ideas*, p.283)

What then is the nature of everyday thinking? It's not sensing!

'Science' is related to thinking. That must be burned into our thinking.

"Everyday thinking is unrefined science." Of course, that isn't how Einstein wrote it. It reverses his "Science is nothing more than a refinement of everyday thinking." If we analyze his statement, though, is it really different?

That statement of Einstein's is more valuable than any other statement he ever made. We should analyze it. Taken literally, it is entirely consistent with the claim that science does not exist. What exist are the everyday thinkings, either primitive or refined, of individual learners like me and you. All of us are learners, at some point in our learning career that started at birth and seems to end with death. Does anyone think there is an actual point in time at which our thinking suddenly turns from unrefined to refined?

Each of us individuals is born belief-less. During the first five or six years of our personal learning careers, we acquire roughly identical sets of fundamental beliefs or convictions. We can refer to them with Einstein's "everyday thinking" or with the phrase "common sense." Even though they form the foundation for all of our later, 'higher' education, we seldom notice how utterly dependent we are on those original beliefs or convictions until they are challenged.

Still, the idea that science is a refinement of everyday thinking is itself a hypothesis. It is a theory. It is a subsidiary part of the larger, very complex theory Einstein created concerning the kind of thinking for which he had become famous.

Another sub-part of his theory about science is his claim that every idea, whether of a kitten or of the speed of light, is a concept created by our imagination:

> I am convinced that even much more is to be asserted: the concepts which arise in our thought and in our linguistic expression are all—when viewed logically—the free creations of thought which cannot inductively be gained from sense experience. This is not so easily noticed only because we have the habit of combining certain concepts and conceptual relations (propositions) so definitely with certain sense experiences that we do not become conscious of the gulf— logically unbridgeable—which separates the world of sensory experiences from the world of concepts and propositions. . (A. Einstein, *Ideas and Opinions*, p.33)

This part of his theory about theorizing is also a theory or hypothesis. His view is that there is a huge difference, a 'gulf' between what we experience on the one hand and the concepts and opinions we use to interpret what we experience on the other hand. This gulf between experience and interpretation is an idea that Einstein repeatedly singled out for emphasis in his explanations of scientific knowledge. But in

21

The Foundation: Every-Day Thinking Habits

1949, just six years before he died, he was emphatic about the necessity of using this theoretical distinction to explain theorizing:

> A basic conceptual distinction, which is a necessary prerequisite of scientific and pre-scientific thinking, is the distinction between "sense impressions" (and the recollection of such) on the one hand and mere ideas on the other. . . . We regard the distinction as a category which we use in order that we might the better find our way in the world of immediate sensations. (A. Einstein, "Reply to Criticisms," p.673. P. Schilpp ed. *Albert Einstein: Philosopher-Scientist*)

This distinction is an unnoticed feature of our everyday thinking. We first experience. Then we seek to understand (or think). Children, upon seeing a strange, unfamiliar object, ask "What is that?" Who of us has not heard an unusual sound and asked "What's that noise?" What sufferer has never asked the doctor "Why does that hurt?" The sensing comes first. What child's first question is "Am I seeing anything?" Who among us always prefaces a what-is-it-? question with "Did I hear anything?" And who visits the doctor to find out whether or not they feel pain? Explanation comes later.

The first recorded thinkers in the West to invent those two distinct ideas and two names to help remember the difference between sensing and thinking were Socrates, Plato, and Aristotle. The distinction seems so common-sensical once it is brought to our notice that it has been referred to by every important thinker who succeeded those long-dead Greeks, even by those who have proposed some other basis for thinking about human thinking. Their distinction between experience and reasoning has become the touchstone for today's myth that great thinkers can be divided into two groups, empiricists who emphasize the priority of experience and rationalists who stress the superiority of reasoning.

Like Einstein who knew better, we must get in the habit of emphasizing the absolute importance of both.

The Wonderful Myth Called Science

Everyday thinking is also distinct from words and things

Nothing is more difficult to think of than the act of thinking and of the thoughts with which we think or understand. Thinking is the ultimate mystery — whether everyday thinking or the most complex imaginable.

But we can begin to notice or be conscious of our thinking if we say what it is not. In part, saying — or rather, thinking — what thinking is not should be viewed as defining it. In that sense, we have already begun defining thinking by saying that it is not sensing. Next to sensing, the things we most often confuse thinking with are language and things. The best way to construct a solid theory about thinking is to become as clear as possible about what we common-sensically, in our everyday thinking, regard as language and what we regard as things.

First, thought is not language. Once we have completed several years of schooling, we find it very natural to connect language with dictionaries. Once we notice or become more fully conscious of what we find in dictionaries, it will help us to understand just why Einstein said everyday thinking is a very difficult thing to think about.

The common, everyday way to think about the words 'making up' language is to think of them as symbols for ideas. If we ask "What does 'vug' mean?", we don't blink an eye if we are told "Look it up in a dictionary." When we do pick up a dictionary with the idea in mind that "Here is a collection of symbols for ideas," we may get a shock. Can there possibly be that many ideas?!

The small, inexpensive dictionary resting on my desk has slightly more than 74,000 entries. That means "myth" should be used for an idea fenced off from the ideas signified by at least 74,000 other words, minus the number of ideas expressible by synonyms. That is a huge number and should prompt every attentive dictionary-user to ask, "Who invented all those words and the concepts they signify?" The full magnitude of that question becomes even clearer if we check further. For instance, one of the other non-foreign language dictionaries in the house has 460,000 entries!

The Foundation: Every-Day Thinking Habits

It is hard to imagine any single person learning all those thousands of words in a single lifetime. On the other hand, most of us have to learn a vastly larger number of words than we habitually realize. It is estimated that average adults use roughly 70,000 of them. Researchers have studied the rate at which normal children learn words. According to them, a normal child learns 14,000 words before their sixth birthday.

> By the age of six the average child has learned 14,000 words. (The estimate of 14,000 words includes inflected and derived words and is based on comprehension vocabulary. For root words only, the estimate falls to around 8,000, or roughly five new root words a day) On the assumption that vocabulary growth does not begin in earnest until the age of eighteen months or so, this works out to an average of about nine new words a day, or almost one per waking hour. So we have a puzzle! Learning even a single new word involves representing a great deal of information, yet the child learns an average of nine words a day. (S. Carey, "The Child as Word Learner," in *Linguistic Theory and Psychological Reality*, ed. M. Halle, J. Bresnan, G. Miller, p.264;) .

Clearly, none of us is the first person in history to invent all of those thousands of ideas and the dictionary's thousands of names for them. On the other hand, every last one of us who wants to understand the ideas of our predecessors must personally invent our own replica-ideas of their ideas to use in interpreting their words. How do we ever learn so many new ideas?

Thinking is distinct from sensing and from words. But are thoughts things? Well, we can begin by saying that, when we are not thinking about thinking, thinking is distinct from the things we think thoughts about. This is especially true if we notice that, in our everyday thinking, we certainly think of thinking as something going on inside us and of the things we think about as things existing outside. We think of stars, but no two things can be farther apart than us and the stars. When we turn our thoughts to turtles, we assume that nothing happens to the stars that are so far away. The 'thought turning' is done inside us.

The Wonderful Myth Called Science

In everyday life, we take it for granted that thoughts are private. If they were not, then there would be no game called "Charades" in which someone desperately tries to give us clues to what she or he is thinking, often without success. That, we say, is just plain common sense.

Scientists, that is the people we usually think are neither philosophers nor theologians, are trained to study things, not thoughts. And, because they do not question their everyday thinking, it never occurs to them to think that the things they are studying are inside them! The things 'scientists' study are — or seem to be — outside them and outside us, i.e., public, not private the way thoughts are.

Comparing thoughts to things gives us a further insight into the difficulty of understanding thinking. There may be a vast number of words in a dictionary, but there are almost infinitely more things in the world than words in a dictionary! Books tell us there are billions of galaxies in the universe. They add that there are billions of stars in each of those billions of galaxies. As for our planet Earth, it was estimated in 2005 to support 6.4 billion individual human inhabitants. If we add the trillions of grains of sand and rocks, plants and trees, insects and animals, the number of real things in the universe is staggering. How can any finite human being even begin to think about all the things that exist?

If you re-read the last sentence, a question, you will notice that you just did think of all the things that exist! The question, however, is not whether we can think of every thing that exists. The question is "How is such a miraculous feat possible?"

Thinking involves grouping

In our everyday thinking, we assume that the universe is 'made up' of more individual things than any of us could even begin to count. Take words, for example. Words make up only a fraction of the things in the universe. But each one of us who is old enough to read and understand the words on these few pages can think (i) about all the words at once, as I just did and you someday will be doing, and (ii) about all the zillions upon zillions of things at once, as I just did and you someday will be doing. (Didn't you do it?)

The Foundation: Every-Day Thinking Habits

Every great thinker who has tried to understand thinking has had to explain how it is that many things can be thought of with a single thought. The answer is that we can mentally group things together. Some grouping is based on shape or form. But we mentally group things for endless other reasons.

First, we mentally group things on the basis shape or form. It is impossible to over-stress or exaggerate how natural this model is. What do we see when we look around us? Things. Trees, squirrels, people. What is the biggest difference between a tree, a squirrel, and a person? Obviously it's shape. If I try to draw what I see, I draw shapes. What is the greatest similarity between all trees? So long as we stick to what we see, it is their shape. We give different names to things with different shapes, and we give the same name to things with similar shapes.

How does this help to understand our knowledge? Start with three individual humans, each with a proper name. Mary isn't Joe, and Joe isn't Sally, and Sally isn't Mary or Joe, etc. But when we say "They are all equally human," we are thinking of all three individuals simultaneously. If we say "Mary is fully human," "Joe is fully human," and "Sally is fully human," we know that we are using "human" to mean essentially the same thing in all three sentences. We can do that because we are using one idea three times, once for each different thought expressed by each different sentence.

Since Mary, Joe, and Sally are all different but the idea signified by "human" is one and the same, Plato hypothesized that "human" is the name for an idea of an essence 'separate' from physical things. Aristotle, though, was more 'down to earth.' He invented a theory that the human mind 'extracts' the same (one?) form from the different flesh-and-blood individuals, and uses it as a mental idea in the mind in order to think about the physical things outside the mind.

The preceding 'draws a rough picture.' Endless details can be found in the library, offering different presentations of Plato's and Aristotle's theory about forms and essences, debates about the theory's details, and books to 'prove' that everything about that theory-model is wrong. It isn't totally wrong! In fact, it can be used to deepen our appreciation of the mystery of knowledge, one of the universe's greatest mysteries of all.

But shape or form is only one basis for grouping. Start again with Mary, Joe, and Sally. We call all three of them "human" or "human being." Why? We also call two of them "female" and one "male." Why? But we may also think of all three as "close friends," in contrast to the billions of other humans. Why? Except in the first case, it is not because of their overall shape. In the second case, only parts of their shapes make two of them female and different from the male. As for the third category, it would be possible to photograph every last feature of the three, but it would be impossible to find anything about their shape or form that makes them similar as friends, but different from the other billions of non-friend humans.

Everyday thinking is highly organized into groups

Why can young students take courses in physics, chemistry, biology, even psychology, without first taking a course in philosophy? For the same reason young students can take a course in philosophy without first taking a course in philosophy.

That 'same reason' is another unnoticed feature of everyday thinking. In this book, "philosophy" will be used as a synonym for "worldview." No one can take a course in anything without already possessing an unimaginably complex worldview or philosophy. *We each have a philosophy by the time we are five or six.* Acquiring it takes place while we are wholly incapable of realizing what is happening. No infant in the crib can think the thought, "I have a lot to learn, but I don't know how I'll ever manage to do it." No one-year-old can think the thought, "I have to learn an average of five new root words every day." And no five-year-old can think the thought, "Wow, I have a worldview, and many of its most important components are group-concepts I never even notice that I'm using."

Why don't sixteen-year-olds and even most sixty-year-olds know they have a worldview, a fantastically intricate set of interconnected beliefs, or a philosophy? It is because most have never been told. Or, if they

27

have, it wasn't often enough for it to really sink in. Nevertheless, noticed or not, everyone old enough has a worldview-philosophy.

To begin 'seeing' how structured everyday thinking is, pause to analyze the game called "Twenty Questions." One player thinks of something. The others must guess what it is. The entire game is based on group concepts. If we can understand the first question, "Is it a person, place, or thing?", it is only because we can mentally divide all the things we know about into three mental groups. If the answer is "a person," we can use sub-groupings to ask whether the person is alive (as opposed to dead), old (as opposed to young), female (as opposed to male), and so on. If the answer is "a thing," we must find out if it is animal, vegetable, or mineral. We can do that only because we can mentally sub-divide every non-person and non-place into one of this second trio of huge, mental pigeonholes. "Animal" names the group-concept made by collecting our ideas of every thing from elephants to chipmunks, the idea labeled "vegetable" contains subsidiary ideas of everything from oak trees to blades of grass, and the idea signified by "mineral" is a bottomless bin for our ideas of whatever is left over.

That common-sense structure or framework forms the foundation for all anti-common-sense views of reality. The one proposed by Descartes in the 1600's was a major step forward for those looking for the ultimate truth about things. Others were proposed afterwards, but they were based on his and his was based on the common-sense structure we use for our 'everyday thinking.'

And that is the bottom line in all of this. No one can begin 'higher' learning in kindergarten or first grade without a largely-ignored, unnoticed, common-sense philosophy. As we will see, some of our original common-sense beliefs are false. Whoever accepts some of the most thoroughly verified discoveries of recent times must surrender some of her or his most firmly-entrenched, common-sense thought-habits. Not all of them, however, only some.

The Wonderful Myth Called Science

Everyday thinking is 'common' to women and men of all time

Before continuing this section on the unnoticed thought-habits which we group together under "common-sense philosophy," one further principle must be mentioned. Every normal human being living on Planet Earth, from the very first one who saw the light of day to those who will be born on the day you read this, can be expected to develop a philosophy *whose most important ingredients are exactly like everyone else's*. When archaeologists speculate about the habits of our primitive ancestors, they assume that those ancestors understood the difference between the places where they hunted and the animals they hunted in those places, between both of those and the place where their village and other members of their tribe were located. The only reason we today can understand what Aristotle or Descartes believed is because they illustrated their advanced ideas with common-sense examples drawn from everyday thinking about persons, places, animals, and so on. Even materialists and idealists can discuss their diametrically opposed world-views by 'translating' their key terms into their closest common-sense meanings. The conclusion here is the same one that Einstein reached in 1936:

> . . . the critical thinking of the physicist cannot possibly be restricted to the examination of the concepts of his own specific field. He cannot proceed without considering critically a much more difficult problem, the problem of analyzing the nature of everyday thinking. (A. Einstein, *Opinions and Ideas*, p.283)

Two further rules about mental grouping and concept creating

In order to prepare ourselves to understand Einstein's theory about the thinking of 'scientists,' we must look further into the difficulties of understanding the everyday thinking that we do so unthinkingly. Two further rules will open the door to more of thought's secrets. Rule #1: We group things in utterly arbitrary ways, that is, in any way our fancy dictates. Arbitrary does not mean irrational. Physicists mentally group things in utterly different ways, but always — they feel — in a way that

29

is reasonable. These arbitrary — and yes, at times irrational — groupings are possible because every thing can be regarded as both similar to and different from every other thing.

Proof for this rule is evident in the ways we actually group things. We create group-ideas for a few things: "that pair of love-birds," "this trio of singers," and "my handful of closest friends." We create group-ideas large enough to include every last thing in existence: "the universe," "the world," "the cosmos," or just plain "reality." We even create group-concepts that embrace things we have no specific knowledge about. Every page in this book will probably illustrate that ability.

This last ability, namely, our power to think about individuals we have no specific, individual knowledge of — except that, if they exist, they are like those we do have ideas for — is mysterious. The Greeks and their successors went to great lengths to explain it. Central to their explanations were complex 'individuation' theories. But 'individuation' theories seem so unsatisfactory, that one expert after another has proposed what he (they were mostly he's in those days) thought was a better one. Whoever desires to study the debates can begin by looking up the name of Zeno of Elea. (His best years were c. 470 B.C.)

In time, the obsession with pinpointing the feature of featureless *thingS* that justifies using the plural rather than the singular was taken to the extreme with the modern theory called "the identity of indiscernibles." It states that, if two things are exactly alike in every imaginable detail, then they are not two, but really one.

Something like that is what led Ockham to revolt. First, we can think of two, two billion, even two zillion things which are identical in every imaginable respect except that each one is distinct from every last one of the others. Second, even if we cannot immediately discover the reason for it, we would be silly to say it is not a fact. That is like saying that, if I cannot explain why I am able to think, I must not be doing it. This book is based on the fact that I can think of a huge number of things whose identifying features, if any exist, I am entirely ignorant of. For instance: this book is written for any and every human being who has acquired a

30

certain amount of knowledge. Were their numbers only a million, it is obvious that I am writing mostly for total strangers. I do not know a million 'proper' names, and even if I did, it would take days to write them out in the opening sentence of this paragraph. This paragraph also refers to this book (not yet written!), but the word "book" (singular and plural) has been used in several earlier paragraphs on several earlier pages. Of the millions of books in circulation, I know the merest handful, but I can think about all of them simultaneously, as I (like you) just did. "Number" names a group or class or category or mental pigeonhole into which it is tempting to say an infinite number of individual numbers of all sorts can fit. "Stranger," "name," "day," "individual," "sentence," "paragraph," "page" . . . How long would the list be, if it included every heterogeneous group named by a noun (proper or common) in this book?

Dictionaries would have thousands more entries if we had to have a separate word for each separate idea of each thing. But we don't, hence they don't. They list several alternative guesses for each entry to help users guess what the writer had in mind. Dictionary" should be viewed as a synonym for "guess list."

So much for rule #1. Equally important is another rule. Rule #2: We can create concepts of both things and groups that do not exist.

"Myth" will be used as a name for concepts of things which do not exist.

"Myth" is a word. We assume it names a concept. We have the word. We have the concept. Do we have a thing as well? Can we have a name and an idea that aren't the name and idea of a thing? Just about the time when we think we are on a roll, understanding matters, we must confront complicating questions like those.

Here are two more. First, we have thoughts about things. But thoughts, are not *they* things, too? Second, we have thoughts about things that do not exist. Can things that do not exist be things?

The Foundation: Every-Day Thinking Habits

Answering those questions requires careful thinking. First, here are the answers. Are thoughts things? In this book, yes. Can things that do not exist be things? In this book, no and yes.

We should be warned that there are many different theories about myths. Individuals who propose them have habits of using "myth" in ways that differ from others' usage habits. In other words, the word "myth" has more than one meaning. Dictionaries give it different definitions. My small dictionary lists three. A larger dictionary offers four choices, but also warns that both the denotation and the connotation of "myth" vary greatly, depending on the views held by the user of the term.

In this book, statements using the term "myth" will generally refer to two different things, namely:

1. to thoughts that are false or

2. to thought-about-things that do not exist.

Thoughts that are false are things that exist; else they could not be thoughts. Every actually real thought is one that is being understood by a really-existing thinker. What makes many thoughts false is that they are 'about' some thing that does not exist.

We should also be warned that there is no agreement whatever on any of the topics discussed in this book. There is vehement disagreement particularly about the topics discussed in this chapter. Obviously, what one person thinks is a true thought, many others will dismiss as false, and what one person thinks is real, many others will reject as mythical. The common-sense view adopted here is that, true or false, thoughts are real, even thoughts of people who believe that thoughts do not exist.

We group thoughts into two types: true ones and false ones. At times, an entire group of false thoughts can be bunched together under "myth," because they are entire stories about things which do not exist, e.g., King Lear, the Star-Trek Klingons, Snow White, etc. The false thoughts are real, but the things they are 'about' are not. The myth-group-concept has a diametrical-opposite, namely, the group-concept for thoughts about things that do exist. Its most common name is "fact." When two people disagree about 'the facts,' each one is calling one or more of the other's thoughts false.

The thesis that our thoughts can be divided into two groups, those which are about things that do not exist, as opposed to those which are about things that do exist, is an underlying principle for this entire book, beginning with its first chapter.

Problems begin when we describe the non-existing objects of the literally false thoughts as myths. When people believe that science is a single, disembodied pool of knowledge rather than individual people's thoughts, or when they believe that the government is an invisible thing over and beyond many individual people plus individual habits plus individual actions, their literally false beliefs go into the mental pigeonhole labeled "myth." But what about the objects, science and government?

Science and governments are not myths the way Santa Claus, Cinderella, Snow White, phlogiston, the ether, and essences are. It is legitimate to deliberately use "science" as habitual shorthand for a group of real thoughts of real individuals and to use "government" for real persons, their real beliefs about their roles, etc. Such habitual uses are not only not objectionable. On the contrary, group-concepts are indispensable as instruments or mental tools needed to think simultaneously about individuals too numerous to talk about one by one.

Another warning. Not every sentence here will explicitly distinguish words, concepts, and real things. For instance, it would more accurate to re-write "Science is a group-concept" as "The word 'science' is the name for a new group-concept that telescopes endless ideas of real people's real thoughts." But it becomes tiresome, unnecessary, and even irksome to repeat the triadic distinction in every sentence. As long as reminders of the critical distinction are inserted often enough, no harm is done by the elision. End of warning.

The Foundation: Every-Day Thinking Habits

Summary: Science is a wonderful myth.

The popular view is that the first scientists, at least in the West, were Greeks such as Thales, Anaximander, Anaximenes, and Pythagoras.. When Pythagoras said he was a *philos* of *sophia*, he was — as they say in literature courses — personifying. There is no wisdom out there, just waiting to be discovered the way Neptune was waiting out there for Galle or Galileo to discover it.

The best way to rephrase what Pythagoras meant or should have meant by his reifying or personifying is to compare it to what we mean when we say "Someone is trying to regain their good health" or "Some overweight and flabby people diet and exercise, and why? Because they are intent on acquiring better health." There is no health out there, just waiting to be acquired. Less healthy people can try to become more healthy the way cool water can be heated. Those in pursuit of wisdom are those who pray to be more truly wise.

Aristotle's famous distinction between substances and attributes should be thought of as a tool for 'reducing' our personifyings or reifyings to more literal beliefs. He 'put into words' what all of us tacitly take for granted, namely, that there are no such things as the—hot, the—cold, the—wet, the—dry out there, only pots and pans and other things to which we attribute the predicates "hot," "cold," etc. Just so, there are people to whom we attribute the adjectives "healthy" and "wise."

That is why "science" as the name for a disembodied 'thing' is a myth. All knowledge is private knowledge belonging to some individual knower. The degree to which a person's personally acquired knowledge is true and is part of a system of other true beliefs — a system broad enough to account for all the relevant evidence — is the degree to which a person's worldview, philosophy, or belief-system deserves to be called "science," that is, a tapestry of tested and true personal opinions.

In the year 2005, we are foolish if we do not each do our best to acquire true knowledge. The years-long efforts of great thinkers, beginning long ago with Thales and reaching down to our own time, have put us, their debtors, in a position to learn more about the world

The Wonderful Myth Called Science

than any of them did. The words, words, words stored in their books are like frigates that can take us wherever we wish to go — even back in time to learn about Thales and his successors. Or, if it is not words in their books that take us where we want to go, it is — or seems to be — whatever those things are that we readers observe. . . .

We readers observe and interpret with our radically diverse mindsets. With what Einstein referred to as 'mentalities.' Mentalities shaped by our different lifetimes of unique personal experiences.

> Furthermore, after some vain efforts, I discovered that the mentality which underlies a few of the essays differs so radically from my own, that I am incapable of saying anything useful about them. (A. Einstein, "Reply to Criticisms." p.665 P. Schilpp ed. *Albert Einstein: Philosopher-Scientist*)

CHAPTER III-I.

Reading: 99% of Scientific Learning

I owe innumerable happy hours to the reading of
Russell's works, something which I cannot say of any
other contemporary scientific writer, with the
exception of Thorstein Veblen.

A. Einstein, *Ideas and Opinions*, p.29

A. THEORETICAL SCHIZOPHRENIA

Common sense versus physics

Stop here, if you are among the faint of heart.

On the other hand, if your goal is to pursue truth open-mindedly,
wherever the trail leads, then read on.

Begin by thinking hard about this. Einstein said that science is a
refinement of everyday thinking. In one sense, it is. In another sense,
however, today's relativity theories and quantum physics are utterly
incompatible with everyday thinking.

Ask yourself, for instance, what you'd think if you were told that,
about five years before he died, Einstein asked someone "Do you think
the moon exists only when you look at it?" Most people, if they believed
that Einstein really did ask such a question, would probably take it no
more seriously than they take the more notorious question, "If a tree falls
in the forest, does it make a sound if there is no one there to hear it?"

Such questions simply make no sense for our everyday thinking.

But those aren't jokes.

You might take those as jokes. (Some Hallmark employee thought up a related question for a recent Shoebox card: "If a tree falls in the forest and then springs back upright as a joke, do the squirrels freak out?") Yet those questions expose the chasm between our everyday thinking and facts about nature that began coming to light at least no later than Copernicus. It was Einstein's keen awareness of that chasm (he had studied Immanuel Kant's radical theories about everyday thinking) that prompted his insistence that we "cannot proceed without considering critically a much more difficult problem, the problem of analyzing the nature of everyday thinking." But for that, it is essential to begin noticing the total conflict between our everyday thinking about hands, books, brains, etc., and most physicists' theories about what those things are really like.

A look at our hand will show what that means. If, when we were born, someone had reached into our crib and lifted one of our hands, we'd have had no idea whatever that the things we saw were hands, that "hand" was the English name for them, or — most importantly of all — that one belonged to us and one didn't.

Like those of the older John Locke (1632-1704), the writings of Jean Piaget (1896-1980) can help us ponder the early stages in our acquisition of everyday thinking. His writings can help us notice the huge step we took in beginning to believe that things we see, e.g., our hands or those of anyone else, do not go out of existence as soon as they go out of sight. They help us notice the equally huge step we took in getting used to the difference between our self and everything else, e.g., between our hands and the hands of other people. But they will never help us complete our education unless we go farther. Much farther!

Contradictory theories about our hands

For instance, we look at one of our hands. As adults, we are as familiar with it as with anything in the universe. It has a color. It is warm. It can be held steady, motionless. If we hold it up to the light, we can't see through it. It is solid.

The Wonderful Myth Called Science

But a *National Geographic* publication, *The Incredible Machine*, informs us that our hand has no color. "There is no red to the rose, no yellow to the bumblebee, no green to the bean. It's all in your head." (P.316). Since our hand is not in our head, it has no color. Later, we read the May 1995 issue of *National Geographic* and marvel at the "Worlds Within the Atom" story. It tells us that "physicists are searching for the ultimate building blocks from which all things—the stars, the earth, you, I, and the atom—are made." (P.634) Atoms, however, are nearly 100% empty space—so much for the solidness of our hand. Many of the atom's building blocks are electrons. J. S. Meyer, in *The ABC of Physics*, astounds us with the claim that "these electrons make approximately 10,000,000,000,000,000 revolutions" around the atom's nucleus every second. (P.35) What happened to our motionless hand? We open to the first page of Max Born's *The Restless Universe* and get our answer. "It is odd to think that there is a word for something which, strictly speaking, does not exist, namely, 'rest'." That, of course, is consistent with Galileo's announcement that nothing physical is warm or hot. We say our hand is "warm," the physicist translates that as "the infinitesimal bodies have a certain amount of kinetic energy, i.e., motion."

Even if we agree with it, however, this 'scientific' view of our hand is largely confined to those intervals when we take a course in introductory physics or watch a PBS *Nova* program. Life outside the classroom, as well as after the TV show ends, goes on the same way it always has. Since we carry our hand with us wherever we go, we can always reassure ourselves that the truth about it is what we've known for as long as we can remember. 'Science' is for the observatory and the laboratory, not for everyday life.

However, the degree to which we believe that both descriptions of what's 'really there' are true is the measure of our easy-to-overlook two-minded-ness. Or should it be easy-to-repress two-mindedness? Arthur Eddington, who led the expedition which gathered the first empirical evidence supporting Einstein's theory of general relativity, was emphatic about the contradiction. On page one of *The Nature of the Physical World* he wrote:

Reading: 99% of Scientific Learning

> When we compare the universe as we had ordinarily preconceived it, the most arresting change is not the rearrangement of space and time by Einstein, but the dissolution of all that we regard as most solid into tiny specks floating in the void. That gives an abrupt jar to those who think that things are more or less what they seem.

". . . an abrupt jar"? Unfortunately, too few of us ever feel this jar, synonymous with cognitive dissonance. (That is an auditory image. Or a mixed metaphor? A musical analogy? If the last, then a member of an aesthetic grouping? See G. U. T. below.)

Eddington's remark is an introduction to the major inconsistency in the worldviews of nearly all of us moderns. The most urgent incompatibility facing every would-be unifying thinker today is not the discrepancy between Einstein's relativity and Bohr's quantum. It is the theoretical schizophrenia, the double-mindedness, of everyone who does not know how to integrate their 'everyday thinking' and their 'scientific' beliefs about material things.

The quest for this latter unification requires open-minded boldness. Thus, if you think you are open-minded, what lies ahead will serve to verify or falsify that facet of your self-concept. If you regard yourself as a bold thinker, ready to follow the facts wherever they lead, what lies ahead will take you on a roller-coaster ride.

But if you're faint of heart, stop now.

B. TO OVERCOME THE SCHIZOPHRENIA

'Science' and the need for a grand unifying theory or G.U.T.

It is a widely held view that by 1925, Einstein's productive career was over. One reason is that none of his later theoretical formulas is as important as, for instance, $E = mc^2$. However, Einstein himself did not feel his contributions were finished. He worked on till the very end of his life in 1955, a fact that prompts us to ask "What prevented Einstein from giving up his 'scientific' pursuits and enjoying a well-deserved retirement in the Sunshine State of Florida?" What was he doing for those last thirty years? What did he think was so important?

The answer is simple. He was doing what Socrates, Plato, Aristotle, Aquinas, Descartes, Berkeley, Kant, and countless others regarded as so important that they, too, worked to the end of their lives. All of them were driven by a desire to understand the universe as a whole. They all sought a unifying worldview. Those who study Einstein's career often use "Grand Unifying Theory" or "Theory of Everything" for what he, like other physicists even today, was pursuing. Those already-shorthand phrases have been further shortened into two acronyms, G. U. T. and T. O. E. Einstein was in pursuit of a G. U. T., a T. O. E., even a G. U. T. O. E. The older shorthand was our familiar English term "Philosophy."

A necessary attitude: confidence that our reasoning can unify

Unifying is an individual affair. Though we depend upon the stimulus provided by reading others' words to learn their alleged discoveries and attempted explanations, it is up to us to select and then piece together the true discoveries and explanations. The reason is obvious. Every alleged discovery — whether it is about phlogiston, the ether, caloric fluid, gravitinoes, or superstrings — is disputed, and every attempted explanation for such discoveries is rejected by at least some experts. Every library-user who pursues a unifying theory must grapple with this overwhelmingly clear truth: there are contradictory views about every alleged discovery and attempted explanation.

How is our personal unifying — the only kind there is — to be achieved? By recognizing contradictions and trying to eliminate them. By using logic. By hard thinking. There is no single term for the process. Reasoning is a good one.

Would-be unifiers must begin with faith in reason. If athletes who train for the Olympics did not hope they might be winners, they would never undergo their rigorous exercises. If those in pursuit of 'the whole truth and nothing but the truth' (not everyone is) had no hope that it could be found, they would have no motivation for this type of serious reading and reflection. From a practical standpoint alone, modern technology shouts a loud message to anyone with ears to hear. Progress is possible!

In Descartes and Einstein, we have two exemplary models. The first, the Father of Modern Science, was certain that truth is possible, and certain he had found it. Einstein, too, was certain that truth is possible, and viewed the increasingly successful search for nature's laws as a thing of wonder:

> The very fact that the totality of our sense experiences is such that by means of thinking (operations with concepts, and the creation and use of definite functions, relations between them, and the coordination of sense experiences to these concepts) it can be put in order, this fact is one which leaves us in awe, but which we will never understand. One may say "the eternal mystery of the world is its comprehensibility." It is one of the great realizations of Immanuel Kant that the postulation of a real external world would be senseless without this comprehensibility. (A. Einstein, *Ideas and Opinions*, p.285)

To those two reasons in favor of confidence in reason, we have the further evidence provided by the great skeptics. David Hume (1711-76) advised that the pursuit of final truth is doomed because it simply exceeds the capacity of our human minds. However, he could not overcome his desire to be certain he was right about that. Relying on his acute mind, he wrote books and abridgments of them, plus lengthy dialogues, to demonstrate with rigorous logic how his skeptical conclusions followed from the premises he selected — with great care! — from the writings of Descartes, Locke, Berkeley, and other discoverers before him. Those writings benefited him only because he read them. What he did in effect was show that certain conclusions reached by earlier thinkers could not possibly be proven the way the earlier thinkers claimed, viz., inductively by direct sense observation. Kant accepted the conclusion Hume so meticulously proved. Einstein emphasized what Hume discovered and what Kant had accepted!

The Wonderful Myth Called Science

A similar tribute to the powers of human reason is found in the arch-skeptic, Friedrich Nietzsche (1844-1900). First, he proved beyond doubt that, if — that is, if — we accept the hypothesis that we are products of a universe that had no prevision of any species' origin and has no mind able to devise anything, now that our species is here, we must accept the rigorously logical theorem that there is no rational basis whatsoever for traditional, common-sense moral principles. Thus, his life, like that of Hume, testified to a supreme trust in the products of his thinking. Rather than imitate beasts, which he confessed have no history, he pondered the history of the ancient Greeks long and hard, then built his new program of good and evil on the truths he was confident he, by his reading, had discovered about Greek history. He spent hours upon hours in his garret, writing down 'his truth' in his books, instead of going outside and 'getting a life.' Many of those who read Nietzsche's words (or translations of them) were 'converted' to his inconsistent views. Einstein, of course, was saddened by the lessons many drew from Nietzsche's German words.

C. LIBRARY BOOKS: INDISPENSABLE RESOURCES

Libraries: repositories of public, collective knowledge?

If science were a public body of collective, growing, self-correcting knowledge, where would it be located? Only one answer even begins to make sense—in libraries—in the books collected in libraries—in the written words and sentences in the books.

Pace Plato, the spoken word is in many respects less valuable than the written word. If the only way for the discoveries made by each generation to be transmitted to later generations was via spoken words, individuals in each new generation would have to be trained to memorize what their elders knew, and be able to help the next wave of memorizers to memorize. But memories are imperfect. And the oceans of detailed knowledge available in 2000 A. D. would require one whole army of memorizers for each 'discipline.'

Plato was right to stress the need for spoken language. Written texts are often difficult to interpret, and teacher-lectures as well as tutorial

43

sessions can lessen the difficulty. But try to imagine a world without writing, books, and libraries.

Bibliographies: proof that reading is the bulk of 'scientific' learning

Few advocates of the 'scientific method' ever notice that almost the totality of scientific learning takes the form of reading. This glaring oversight produces some delicious ironies.

For instance, the vast majority of psychology researchers regard themselves as scientists using the scientific method to ascertain their bodies of truth. B. F. Skinner, in 1953 devoted considerable effort to explaining the nature of science. Among his observations was this:

> Science is first of all a set of attitudes. It is a disposition to deal with the facts rather than with what someone has said about them. Rejection of authority was the theme of the revival of learning, when men dedicated themselves to the study of "nature, not books." Science rejects even its own authorities when they interfere with the observation of nature. (B. F. Skinner, *Science and Human Behavior*, p.12)

The absurdity of that claim is easy to expose. Examine an average introductory psychology textbook. For instance, one edition of an oft-printed (scientific) text began to explain the nature of science with this paragraph:

> The essence of science, from the standpoint of the scientist himself, is "a disposition to deal with the facts rather than with what someone has said about them." This means that the scientist must observe for himself and that what he observes is the primary basis of his speculations and conclusions. The importance of observation for science is brought to a focus in the following story from Francis Bacon, a leader (1605) in the history of scientific investigation. (N. L. Munn, *Psychology: the Fundamentals of Human Adjustment,* 5th ed., p.5)

The Wonderful Myth Called Science

The quotation is, of course, from Skinner's book. Without books, Francis Bacon's story would never have come to Munn's attention. Most of all, the silent witness to nearly all of the higher-learning or scientific observation Munn ever did comes at the end of his text. There are forty-one pages of footnotes, followed by an index that lists more than 1300 names of authors whose works Munn found useful — the index to Skinner's book tells the same tale. If Munn or any author of a scientific psychology text had been born in a jungle village inhabited by aborigines and, when they were old enough, had been asked to create an up-to-date psychology text with no years of text-based courses and no access to libraries full of written sources, what might we expect? Perhaps anthropologists can help here?

Catching-up learning

Of course, new experiments and novel observations are part of the history of theoretical progress in understanding nature. Popularized histories of the advances in astronomy, physics, chemistry, biology, etc., focus on them. And, because their authors use analogies from everyday common-sense thinking, they serve as valuable bridges from our original philosophy to otherwise esoteric theories.

That, for instance, is what makes Martin Gardner's the best popularized introduction to the bizarre features of Einstein's relativity theories. Not only does Gardner understand those theories. He also has a flair for apt, everyday analogies. And his *Relativity for the Million* or *The Relativity Explosion* (the later edition) is full of pictures, far more than Einstein's own popularization. No one but a physicist would ever say that his is not better than Einstein's.

But there is one all-important fact gets lost in these stories of scientific progress. That all-important fact is mentioned 'in passing' by Descartes in Part Six of his *Discourse on Reason*. After noting that each individual's learning is limited by the relative shortness of life and the narrowness of his or her personal experience, Descartes continued as follows:

Reading: 99% of Scientific Learning

I judged that there was no better provision against these two impediments than faithfully to communicate to the public the little which I should myself have discovered, and to beg all well-inclined persons to proceed further by contributing, each one according to his own inclination and ability, to the experiments which must be made, and then to communicate to the public all the things which they might discover, in order that the last should commence where the preceding had left off; and thus, by joining together the lives and labours of many, we should collectively proceed much farther than any one in particular could succeed in doing. (R. Descartes, *Discourse on Method*, trans. Haldane & Ross, I:120)

". . . in order that the last should commence where the preceding had left off." No individual learner, born as belief-less as all of us are, can begin where anyone before us has left off. We must first spend five or six years acquiring the common-sense foundation for all higher learning. Only then can we go to school, understand lectures and read texts, thereby converting parts of the common-sense foundation into knowledge advanced enough to begin doing experimental research.

The validity of this book's hypothesis about the wonderful myth called "science" depends on two major innovations. First, its attention to everyone's common-sense philosophy or worldview as the foundation for all later learning. Second, its adoption of William James' best description of thought itself.

Reading as the major form of 'catching up'

The fact that reading 'words' is the dominant kind of observation needed to achieve scientific knowledge about the world is perhaps most easily seen in the case of astronomy. Where would Ptolemy (90-168) have been without Euclid's texts and the records of earlier astrologers who plotted the constellations in the zodiac that helped to spot the 'wandering stars' we now call "planets"?

What could Kepler (1571-1630) have discovered without Copernicus' text as well as the more accurate records of Tycho Brahe's star-charts?

The Wonderful Myth Called Science

What could Newton (1642-1727) have achieved, had he not had access to texts which 'passed on' to him the theories of Copernicus, Galileo, Descartes, and especially Kepler? Why especially Kepler?—because his patient mathematical calculating and re-calculating (without any pocket calculator!) was enough to break a centuries old tradition, a veritable logjam preventing progress in the thinking of astronomers. Where he and his predecessors had 'seen' only circles, circles, and more circles when they gazed into the sky, Kepler discovered that they should begin thinking ellipses, ellipses, and more ellipses. That allowed Newton to integrate celestial and terrestrial physics. It is doubtful anyone today would know his name, had he not had access to the discoveries of Kepler who died before Newton was born.

Recall that Munn wrote, "The scientist must observe for himself" and that "what he observes is the primary basis of his speculations and conclusions." Until future learners begin to think "reading, reading, and more reading" when they see the phrase "scientific observation," the same type of logjam will prevent their escape from the myth called "science." and from the myths called "philosophy," "theology," etc.

D. SEEKING THE WHOLE TRUTH IS AN INDIVIDUAL TASK

(Re)attaching the 'sciences' (small s and plural) to everyday thinking.

René Descartes (1596-1650) used a very long title for his 1637 masterpiece. An English translation of it is *Discourse on the Method for Conducting One's Reason Well and for Seeking the Truth in the Sciences.* Distinguishing 'Science' (capital S) as a group concept from 'the sciences' (small s) as its members offers an easy way to correct some bad thought-habits contracted in high school and college.

Step one of the correcting begins with the lesson that, if science of any kind exists, it is thought or knowledge. The only thought or knowledge we have direct empirical evidence for is that which has been acquired by some individual woman or man. And the first knowledge acquired is always the common-sense worldview or philosophy we use for our everyday thinking.

The correcting continues with the lesson that every common-sense worldview or overall belief-system already includes concepts related to the 'special sciences.' For instance, every child who understands "Twinkle, twinkle, little star" does so only because she or he knows something about the pinpoint lights visible in the night sky. Her or his knowledge of stars is framed or de-fined by the other concepts of a common-sense worldview, namely, concepts of persons, places, living things, vegetables, sun, moon, day, night, sky, earth, and so on. Every expert astrologer and astronomer who ever lived built her or his specialty on her or his early common-sense ideas.

Note well! That means that Ptolemy, Copernicus, and Kepler did not begin as non-astronomers and only later begin to be astronomers. By the age of five, they were already budding astronomers. In their prime, they had vastly expanded their early astronomy and become better astronomers.

Step three recognizes that the difference between anyone's first philosophy and their later expertise parallels that between, say, Kepler the child astronomer and Kepler the brilliant discoverer. By the age of five, every normal learner is already a budding physicist, chemist, biologist, and psychologist. That includes every aborigine born at any distant moment in time, in any jungle or savannah.

"Physics," "biology," etc., name fiction-type group concepts

We all get in the unchallenged habit of saying we are taking courses in psychology or studying astronomy. Unfortunately, we are never forced to develop the habit of thinking clearly about the actual facts. Yet, once we are in the habit of distinguishing thoughts in our mind from things outside, the facts are easy to grasp.

The difference between learning about stars and learning about astronomy is the difference between learning about real things that have existed for eons and learning about humans' beliefs about stars. The stars

The Wonderful Myth Called Science

twinkle in the sky, not down here on earth. Astronomy exists in the minds of humans down here on earth, not up in the sky. But, like science in general, philosophy in general, or theology in general, astronomy in general does not exist. Believing in it is belief in a myth.

The only right way to think of studying 'astronomy in general' (singular) is to *pretend* that we extract from each human's mind just her or his ideas about certain or particular parts of the universe, group them into a single, huge pile, and call that imaginary pile "astronomy in general." But the pile does not exist, only the real individual people, their individual beliefs, and the things (e.g., stars) their beliefs are about.

Unless and until we reattach our ideas of 'astronomy,' 'physics,' and so on, to our everyday thinking in that way, we will be like children who parrot what they hear 'everyone' say but who really do not know what they are talking about. William James advocated memorizing teachers' and writers' words. (After all, we need to pass our courses.) But he warned against memorization devoid of well-grounded understanding.

> I go back to what I said awhile ago apropos of verbal memorizing. The more accurately words are learned, the better, but only if the teacher make sure that what they signify is also understood. It is the failure of this latter condition, in so much of the old-fashioned recitation, that has caused that reaction against 'parrot-like reproduction' that we are so familiar with today. A friend of mine, visiting a school, was asked to examine a young class in geography. Glancing at the book, she said: "Suppose you should dig a hole in the ground, hundreds of feet deep, how should you find it at the bottom-warmer or colder than on top?" None of the class replying, the teacher said: "I'm sure they know, but I think you don't ask the question quite rightly. Let me try." So, taking the book she asked: "In what condition is the interior of the globe?" and received the immediate answer from half the class at once: "The interior of the globe is in a condition of igneous fusion." (W. James, *Talks to Teachers*, pp. 105-06)

Reading: 99% of Scientific Learning

Science (capital S): a revision of common-sense

Descartes gave another long title to his greatest masterpiece of all. It was published as *Meditations on First Philosophy in which the Existence of God and the Distinction between the Soul and the Body Are Demonstrated*. The important part of the title is the beginning, which justifies the custom of calling his 1641 work simply *Meditations on First Philosophy*. His aim was to show that, in order to achieve certainty about the realities each of the various sciences (in the plural) are supposedly related to, it is necessary to radically revise parts of our original worldview.

In a word, some of the pillars of our everyday thinking — our First Philosophy — are radically false. Consequently, many facets of our youthful astronomy, physics, chemistry, and so on, are unreliable. Unless we grasp what Descartes set out to accomplish with his 1641 masterpiece, we will never understand what Einstein had in mind. Recall the words of Einstein:

> . . . the critical thinking of the physicist cannot possibly be restricted to the examination of the concepts of his own specific field. He cannot proceed without considering critically a much more difficult problem, the problem of analyzing the nature of everyday thinking. (A. Einstein, *Opinions and Ideas*, p.283)

Science is a revision of our First Metaphysics

If Descartes' 1637 masterpiece focused on the sciences, his *Meditations on First Philosophy* in 1641 deals with the single foundation for all of them. The two works go hand in hand. "The sciences" refers to just portions of our everyday view of things. "First Philosophy" is best regarded as referring to the worldview-framework for the 'sciences.'

At first, our personal sciences (small s and plural) appear coherent. They fit together into our original, twenty-question framework which is our original Science with a capital S. It is our original Philosophy with a capital P. It is our original Belief-system with a capital B. It is our original Worldview with a capital W. In order to fit what Descartes aimed at in his 1641 work into the re-organizational framework presented here, common sense should be called our original Metaphysics with a capital M. What Aristotle and Descartes meant by "First Philosophy" should be translated into the vocabulary of those whose goal is a Grand Unifying Theory.

Our original Grand Unifying Theory, common sense, has been tried by recent great thinkers and found wanting. It no longer 'cuts the mustard' for those who desire to integrate all of the great discoveries of the past four centuries.

As Descartes and Einstein both realized, we need a new and better Grand Unifying Theory which incorporates the best of common sense, but is better as a whole than the unrevised common sense we use for our everyday thinking.

Each of us is on our own

We are born one by one. Even twins are. We die one by one. Between birth and death, we learn one by one. Even when we are reading copies of the 'same' book, we are reading one by one, that is, individually.

That is the premise of this work. A major conclusion is that no one of us presently knows anything that she or he has not personally learned and not forgotten. If there is any other kind of knowledge over and beyond the personal knowledges (in the plural) of individual persons (in the plural), there is no agreement about it. Moreover, any opinion about an alleged im-personal knowledge would be just one more person's personal opinion. There are no thoughts floating about in splendid isolation.

Fortunately, unless our experience changes, it will always seem that we have to make our own discoveries, even when they are our own discoveries of what others discovered before us. Why fortunately? Because part of the thrill of learning is the feeling that we are not mere passive lapdogs. What others have discovered before us can assist us in making our own thrilling discoveries, even if we are not 'historically' the first to do the discovering.

If today's first-time visitors to the Grand Canyon could not thrill to the awesome sight the way thousands of earlier first-time visitors did, the bears, beavers, and bugs would have it all to themselves.

The division of labor: we get to choose our own experts.

This world is so full of details that none of us can be knowledgeable about more than a relative handful of them. That is why there is a division of labor among experts. Unfortunately, as you will see, experts disagree with each other. Which experts know the truth? Experts disagree about that, too.

It is said that Catholics 'shop around' till they find a confessor whose views on what is or isn't sinful match their own. True or not, that is certainly true about us in our search for the experts who know the truth, and each of us must take full responsibility for our own shopping around. Even when we take the advice of an expert regarding which experts to follow, we have full responsibility for choosing the expert on experts.

Throughout these pages, you will discover which authorities regarding which topics — in my opinion — know what they are talking about. But, if the devil can quote Scripture, I can quote Einstein. If that doesn't make the devil right, this doesn't make Einstein (or me) right. You'll have to decide on each topic.

E. READING: A PERFECT SCIENTIFIC TEST OF THE SCIENTIFIC METHOD

The title of this chapter introduces a major theme of this book. If you can unravel the complexities of what you are doing right now, that is reading, you will have taken a giant step toward learning about higher, post-childhood, and post-common-sense philosophy. Moreover, by learning to think precisely about what you are seeing as you read, you will come to see that reading constitutes the ultimate experimental test of 'the' scientific method.

Wait until you see what Einstein thought about Russell's views on sensing!

CHAPTER III-II.

Reading: The Scientific-Method Paradigm

This is not so easily noticed only because we have the habit of combining certain concepts and conceptual relations (propositions) so definitely with certain sense experiences that we do not become conscious of the gulf—logically unbridgeable—which separates the world of sensory experiences from the world of concepts and propositions.

A. Einstein, *Ideas and Opinions*, p.33

A. OCKHAM'S RAZOR

If there were a scientific method . . .

The myth that there is a public body of knowledge distinct from philosophy, theology, and pre-scientific common sense, is twin to the myth that there is a special method for acquiring that knowledge.

But if "science" is taken as a group label for individuals' personal opinions which are true, logically coherent, and all-inclusive enough to account for all of the relevant evidence, then it might make sense to search for a special method to achieve that knowledge. But what would it be?

Sherlock Holmes put his mental finger on part of the answer when he told Watson, "It's an old maxim of mine that when you have excluded the impossible, whatever remains, however improbable, must be the truth." Half of being confident we have the true answer to a question is knowing we have eliminated every last one of the impossible alternatives.

Reading: The Scientific-Method Paradigm

Ockham's Razor

Einstein had a similar elimination idea. Sherlock's rule is implicit in the description of science's aim which he included in a 1936 lecture entitled "Physics and Reality." The formula was used to introduce Chapter I.

> The aim of science is, on the one hand, a comprehension, as complete as possible, of the connection between the sense experiences in their totality, and, on the other hand, the accomplishment of this aim by the use of a minimum of primary concepts and relations. (A. Einstein, *Ideas and Opinions,* p.286)

". . . by the use of a minimum of primary concepts and relations." As Chapter I explained, that phrase relates to the principle picturesquely described as Ockham's Razor: Do not multiply entities without necessity. The 'razor' part of the principle means, "Get rid of all superfluous clutter in your theory." Chapter I applied the 'razor' to theories about science, philosophy, and other collective bodies of public knowledge. By eliminating the entanglements of those myths, the truth stands out with greater clarity:

If, therefore, we begin taking what we hear and read about science and asking "What real things are being talked or written about?", it becomes apparent that, with few if any exceptions, those real things are human beings and their opinions, things we already know about and believe in. (Chapter I, above)

Naturally, experts have completely different opinions about which entities are necessary and which are not. What seems impossible to one person seems not only possible but self-evident to someone else.

Libraries: from science to books to words

It was noted earlier that, if collective bodies of knowledge, such as science, did exist, libraries would be the place to look for them. Suppose, therefore, that you were asked which of the world's libraries you would like to own. If you can even imagine having such a desire, what criterion would be uppermost in making your choice? Would you want the library with the most old classics unavailable anywhere else? One with the most biographies? Physics texts? Novels?

Not one of those criteria comes even close to being the most important. No library books will be immediately useful unless you can read them. For that, they must be in a language you can understand. For most of us, that instantly eliminates the vast majority of the world's libraries.

But even if our library's books are 'in a familiar language,' that isn't proof that knowledge itself is contained in them. We think we see words and sentences in the books. We don't. Words are like collective science: Both are mythical.

Here, "primary concepts" means concepts of things that really exist.

Einstein's phrase, ". . . by the use of a minimum of primary concepts and relations," is part of a lengthy and complex lecture about the relation of physics, i.e., a system of concept-using beliefs, to reality, i.e., *to things that exist independently of beliefs or theories*. It is clear from his lecture that he uses "primary concepts" to mean ideas that are of major importance in the belief-system. It is also clear that Einstein was keenly aware of the fact that there have been and still are conflicting theories labeled "physics," which use different primary concepts. For instance, the concept of continuous fields was basic to his relativity physics, whereas the concept of discontinuous 'quanta' dominated theories about quantum physics. These two cases are examples of theoretical disagreements about what really exists.

Implicit in Einstein's thinking was the common-sense notion of truth. As he explains in the first chapter of his *Relativity: the Special and General Theory*, only theories that have a 'correspondence' to reality are regarded as true.

55

Reading: The Scientific-Method Paradigm

How is 'correspondence' to be determined? Einstein's answer was always the same. Experience, sensory experience, sense perception were terms he used to designate the touchstone for true conceptual theory.

But we will always need subsidiary, 'false' concepts.

Chapter I explained how it happened that, during previous centuries, people invented such fictitious concepts as science, philosophy, and theology (religion) in order to build theories about which knowledge is the best. Chapter II explained our everyday-thinking need for group concepts to help refer simultaneously to things too numerous to think of one by one. Both chapters referred to 'false' ideas, such as that of Santa Claus, Darth Vader, and others, used for purposes of entertainment.

Modern sciences have led to the creation of an endless number of similar concepts for which there literally are no exact counterparts in reality. These have come to be known by many names, ranging from "theoretical construct" to "logical fiction." These constructed or created, subsidiary concepts can be helpful, even indispensable, in the quest for true knowledge. They can also be huge obstacles that impede that quest. Each of us must decide for ourselves which fictions are which.

The greatest errors come from not correctly drawing the line between which concepts give us a grasp of matching realities and which are useful fictions. But by correctly organizing our own philosophy or worldview in such a way that (i) we have true beliefs about things which really exist and also (ii) mastering theories which are false but useful, even indispensable, we can achieve a belief-system that deserves to be called "scientific." That is how Ockham's Razor should be applied, viz., with an eye on this crucial distinction.

The aim of this segment of Chapter III

As the title of this sub-chapter indicates, the focus will again be on reading. It will explain why the correct analysis of what you are doing right now is precisely what everyone does who conducts any kind of 'scientific-experiment' observation.

The Wonderful Myth Called Science

As you read, you will be shown how to 'reduce' all language-related concepts — from ideas of letters to ideas of books — to the status of indispensable logical or pragmatic fictions.

There are two reasons for calling the analysis of reading "the scientific-method paradigm." First, according to Einstein, none of our concepts are derived from sense experience. In no case is that easier to show than in the case of language-related concepts. Secondly, few concepts are as essential as the language fictions for discovering the true nature of thought, which is the essence of all knowledge, no matter what label is projected onto it.

A footnote regarding 'the' scientific method

Whoever is interested in the history of theories about 'the' scientific method will find ample resources in any well-furnished library. Plato proposed dialectic as the method, at least when he was not advocating Socrates' method of definition or the method of classifying. Aristotle wrote six books on 'the' scientific method; when they were collected, they were called *The Organon*. Bacon is famous for his *Novum Organon*, Latin for The New Organon. Descartes wrote A *Discourse on Method* to give his opinion on the subject. *The 24 Science Wars* lectures by S. L. Goldman (The Teaching Company,) and *What is this thing called Science?*, by A. F. Chalmers will explain why there are now more contradictory theories about the 'true' method of science than there are contradictory theories about gravity!

Only William James got it partly right, ". . . the truth remains that, after adolescence has begun, 'words, words, words' must constitute a large part, and an always larger part as life advances, of what the human being has to learn." Only partly, because it begins well before adolescence!

B. WHAT DO READERS SEE?

Notice that sense impressions and thoughts are distinct.

If there is one essential prerequisite for recognizing how much of today's thinking about 'science' is myth rather than fact, it is noticing the difference between observing and thinking. (See Chapter II.) Both sensory experience and conceptual thinking occur while we are reading, but they occur so simultaneously that we 'do not become conscious,' i.e., do not notice, the difference. But the distinction, crucial to Einstein, is found in all thinking, whether everyday or scientific thinking.

> A basic conceptual distinction, which is a necessary prerequisite of scientific and pre-scientific thinking, is the distinction between "sense impressions" (and the recollection of such) on the one hand and mere ideas on the other. . . . We regard the distinction as a category which we use in order that we might the better find our way in the world of immediate sensations. (A. Einstein, "Reply to Criticisms," p.673)

With that distinction in mind, we ask a simple question: Has anyone ever seen — sensed — a number?

Has anyone ever seen a number?

Look at *5*. We all immediately think it is a number. That is only because we all have a long-standing habit of associating (traditional) or "combining" (Einstein) the idea of the number five with what we see when we direct our gaze at 5. We think we see what we are in the habit of thinking about when we see 5.

Now look at *V*. Is that a letter, a capitalized form of a small v? Or is it the number five as the Romans might have said? Look next at *101*. You might think it is the number one-hundred-and-one. But, if you are familiar with the binary code used by computer programmers, you will recognize it as the number five. But, if all 3 of those are 'the' fifth number, it raises a problem: "How many 5's are there?"

The Wonderful Myth Called Science

The solution is the thought that *5, V, 101*, and *five*, so visibly not the same, are not the real number itself. No one has ever seen a real number. That is why Einstein used numbers (integers) to support his claim that all of our ideas are created. They are not mental copies of what we have sensed.

> This is not so easily noticed only because we have the habit of combining certain concepts and conceptual relations (propositions) so definitely with certain sense experiences that we do not become conscious of the gulf—logically unbridgeable—which separates the world of sensory experiences from the world of concepts and propositions.
>
> Thus, for example, the series of integers is obviously an invention of the human mind, a self-created tool which simplifies the ordering of certain sensory experiences. But there is no way in which this concept could be made to grow, as it were, directly out of sense experiences. (A. Einstein, *Ideas and Opinions*, p.33)

Einstein is emphatic. We do not get ideas of numbers by seeing them. We create those ideas. We do it while we are still at the stage of pre-scientific thinking. We begin by learning to repeat what others say. That should recall the claim in Chapter I that we develop our concepts of science, philosophy, theology (and religion), etc., from hearing the 'words' used by others.

But, if what we see on the pages of books are not numbers, then what are *5, V, 101, five*, and so on? What we see is not nothing, so what is it that we see?

Has anyone ever seen a word?

Words. Are they any more visible than numbers? Or are they the same, that is, invisible? We begin our search for the truth by using the word same the same way we used the number five. Try to think with absolute precision about what you see when you look at the word same.

Reading: The Scientific-Method Paradigm

Begin with the fact that we use same to mean totally un-same things. For example, ten glasses that are the same can fit on the same shelf. Analyze that. First, the second half of the sentence means that they can fit on one shelf. In that context, same is used as a synonym or substitute for "one." Second, it would make no sense to say that the ten glasses are one. The first half of the sentence means that the ten glasses are alike or similar. Same is used there as a synonym for "similar," and similarity can range from being exactly alike to being only roughly similar. (Recall the discussion in Chapter II of Rule #1 vis-à-vis mental groupings.)

Now, look again at the preceding paragraph. We might say that one word, "same," appeared five (or 5 1/2) times. But is that possible? Besides the sentences which have a word or words that look exactly like same in them, is there also an invisible word same that is visibly copied each time the four letters s-a- m-e are strung together? Or is there a nearly infinite number of sames? Perhaps both, one invisible word plus innumerable visible reminders for it? (See Plato.)

Part of the solution: be decisive!

One of the most important rules for thinking with precision is this. When it is crucial, be decisive the way Einstein was. When it is crucial, practice making your talk precise enough to fit your thought or belief.

That is the opposite of defending your old thought-habits when they ought to be changed. There is nothing wrong with talking about the sun rising and the sun setting, or about our hand being colored, warm, and solid. But, if Copernicus and Rutherford had been content with their old, everyday-thinking thought-habits, we would not have the heliocentric and atomic theories. It is essential that we resist the various half-measure defenses we are all tempted to fall back on when our present way of thinking is confronted with serious challenges.

For instance, 'symbol' is a half-measure often used with relation to *5, V, 101,* etc. We say those are visible symbols or signs for the invisible number five. We might also say that s-a-m-e is the visible symbol for the invisible word or idea. But what is a symbol? Is symbol any different from same? If "symbol" is a word and words are symbols for ideas, then symbol is a visible symbol for the invisible word 'symbol.' Does anyone think that, when they look at symbol, they see two things, a word and a symbol? Or that, when they look at 5, they see a number and a symbol?

Here, then, is the importance of Ockham's Razor. Look at *5, same,* and *symbol.* Obviously, we can mentally group them as members of the class or set labeled "symbol," but there is no more visible similarity to what we see — 5, same, symbol — than there is to Mary, Joe, and Sally. (Chapter II.) And what is a class, a set, or a mental group? Being decisive means admitting that, though concepts of those fictions exist and are useful, the 'things' themselves — numbers, words, classes, sets, groups, etc. — do not. They are mythical or fictitious entities. Mental fictions. Like Santa Claus, they are thought-about things that do not exist. We must be bold in dealing with theories about numbers, words, etc.

The preceding is one of the dozens of hypotheses being proposed to you as you read this book. Each of them leans on the others. None of them stands in splendid isolation. Hypotheses are thoughts, the only thoughts that exist belong to some individual thinker, and none of them exist in splendid isolation from the rest of the individual thinker's system of beliefs.

Once more, then, be prepared to make your talk fit your thought and not vice-versa. That requires being alert when using two or more different names for one (type of) thing or one name for different things. In everyday thinking, the rule can be violated with impunity. We can use a bat to kill a bat. A ruler can be used to measure the length of the ruler's foot. I've eaten the same left-overs for the last three days. And, yes, ten glasses that are the same can fit on the same shelf. But when precision and consistency are demanded, violating the rule is disastrous.

C. AT MOST, READERS SEE INK MARKS

Shaped and arranged ink marks

If what we see when we look at *5, same, and symbol* are neither numbers nor words, then what do we readers see? This is a crucially

Reading: The Scientific-Method Paradigm

important question for the thesis that "Science" is a symbol for a wonderful myth.

Yes, symbol. "Symbol" can be used to evoke an important idea or concept. There are no fictions as important as language-fictions for learning what higher learning is, regardless of whether that learning begins with Sesame Street programs, Montessori classes, kindergarten instructions, or first grade lessons. That is, there is no whole set of fictions more essential for learning how we do our higher learning than language fictions. What seem to be words but aren't, are indispensable for discovering something invisible and intangible, thought itself.

What seem to be words but are not words, are nevertheless sensible things. Where are those sensible things? In our everyday thinking, we think they are on the pages of books stored in the library. For the sake of argument, pretend that books and libraries exist and can be seen. Then weigh the evidence for this hypothesis: What readers see are specially-shaped ink marks applied by a printing machine to pieces of white paper. The ink marks have an extremely limited variety of special shapes. Each distinct ink mark is similar to numerous others.

Besides their different shapes, the ink marks can be arranged in different ways. An example that combines different shapes and arrangements is: "GOD" and "DOG." Those six ink marks, each located on a different part of the paper's surface, will evoke radically different ideas, just the way clouds do for cloud-gazers, and streaks on a photograph do for viewers. Whoever believes they see more than differently shaped and differently arranged (dried) ink marks on a piece of paper that originally came, neat and clean, from a paper mill, is not noticing the difference between seeing and thinking.

Ostensive definitions

Many twentieth–century thinkers distinguished (i) words whose meaning can be conveyed by dictionaries (which use words to tell what other words mean) from (ii) words whose meaning can be taught only by showing someone the thing that the word is the name for. In case #ii, the learner must see what is pointed to and mentally connect what they see to the word-sound he or she must hear.

That proposal is only a half-truth. First, hold up your hand, look at it, and ask yourself "What do I see?" Or, just hold up your hand and ask a friend "What do you see?" Unless you or the friend are in extremely rare situations, the full answer will be a long one. Besides your hand, you will see far more than just your hand. The only satisfactory short answer to nearly every "What do you see?" will be "A total visual field or TVF." (TVF is an easy-to-remember acronym that can serve as a shortened form of a short answer.) Only individuals already old enough to have acquired a common-sense worldview can begin to learn the meaning of additional native-language or foreign-language words by the showing or ostensive method.

Secondly, we not only see many things at once, we also have many concepts habitually associated with each of them. Try 'the R E D experiment.'

R E D

Imagine a child who is well along in her course of 'lower education.' She has already acquired a common-sense — person—place—thing, — animal—vegetable—mineral worldview, as well as a basic set of English-language words for the items in that worldview. Suppose you want to introduce her to 'higher education.' What would you teach her to call what is between this and the previous paragraph? To be certain she knows what you are referring to, imagine yourself pointing at R E D. Here are a few possible names or descriptions for concepts of what she and you see: Three ink marks. Three shaped ink marks arranged in an order which is the reverse of D E R., three letters, the eighteenth, fifth, and fourth letters of the alphabet, three symbols, three symbols for three different sounds, three symbols for one complete sound, a syllable, the second syllable of sacred?, a word, a name, black, black things, black figures against a white ground, the name for the color of blood, small things (in contrast to the Empire State Building), huge things (in comparison with a subatomic particle), a morpheme, an acronym,

63

something visible, what you see, what you're pointing at, what is equidistant from the sides of the page, what is at the center of her TVF and yours, an example of things which are physical, a misspelling of reed, the answer to "What color is a cardinal?"

Which of those is the most precise, scientific answer to "What do you see there?" The answer is obvious. Depending on the context, each of those answers might be useful. In the present context, however, Ockham's Razor dictates that "shaped, arranged ink marks" is the most precise of all. (So far.)

The evidence for this hypothesis is overwhelming in its sheer volume. If there were a stack of books, each using a different 'genuine' alphabet or script (think of Chinese, Japanese, Egyptian hieroglyphics, Mayan glyphs, etc.), and they were mixed with books filled with 'fake' script, who could instantly identify which are genuine words and which are fakes? Even a laboratory inspection with a microscope will not reveal anything on the pages except shaped, arranged ink marks. Whoever sees more is seeing 'ghosts in the ink marks.'

The bold correctness of Einstein's premise is far-reaching. Here is a sample of language-related concepts of things never-observed: letters, vowels, consonants, syllables, prefixes, suffixes, clauses, phrases, sentences, paragraphs, chapters, novels, essays, prose, poetry, grammar, spelling, meaning, etc. Have you ever noticed that a is a letter, a vowel, an article, a word, and at times the subject of a sentence?

Precise thinking and "as such"

It comes as a shock to most of us when we learn that, though mathematicians all agree that "1 + 1 = 2" is true, they disagree if asked "What are you talking about?" Some argue that there really are invisible, intangible, and immaterial numbers. Einstein argued that we have concepts of numbers, but he didn't believe in numbers as such. Others believe that only the visible symbols themselves exist, that they can be used in various ways when we interact with real things, but that none of the real things are numbers as such.

The Wonderful Myth Called Science

"As such." That expression is useful when we wish to say that something is or is not a real thing distinct from every other real thing. We can refer to the corpse in the coffin as "Uncle Charlie." We might say that Uncle Charlie looks so peaceful, much better even than when we visited him a week ago. But we know that is not Uncle Charlie as such. What is there is a corpse as such. Or we might say we saw Mommy kissing Santa Claus under the mistletoe, even though we know it was not Santa as such because it was Daddy as such.

One way to think precisely about "What do readers see?" is to count the things that are being talked or, rather, thought about. Unless people believe that both the dead corpse (a redundant phrase) and Uncle Charlie exist, or that both Santa Claus and Daddy exist, then in each case there is only one entity in existence, however many different thoughts we can think about it, that is, about what exists as such. Similarly, readers who are not victims of double-vision do not see numbers or words as such, only ink marks as such. Thinking that we see ink marks and words is like thinking that the stranger who just stepped off a plane is both a man and our nephew.

This idea of counting helps to understand why scientific thinkers must acquire the habit of thinking 'with mathematical precision.'

D. THINKING WITH PRECISION ABOUT OBSERVATION

Scientific observation: who does it?

Who does it? Only one answer is true: Everyone must do his or her own. Genuinely critical readers will regard large sections of well-equipped libraries as a silent cacophony of debaters. By reading their books, we discover that, like the mathematicians, the experts in every meadow (variant of "field") disagree on the bottom-line assumptions in their respective disciplines (variant of meadow). Unlike jurors who must master only two contending theories regarding the defendant (proven or not proven guilty), critical thinkers in pursuit of a unified, up-to-date worldview must master opposing views on dozens of basic issues. And then critically choose the right view on each one.

Reading: The Scientific-Method Paradigm

And one of the most basic of all primary issues is the question, "What do scientists observe?" Since the 'higher education' of most of scientists comes from reading, most of what they observe will be the words and sentences they scan or ink marks they see. The implication of that overlooked fact is earth-shaking. What others have personally observed is never a personal observation of our own!

It is a blunder to imagine that another's observation is our own.

We can delude ourselves with the thought that most of our 'scientific knowledge' is based on observations made by others in observatories and laboratories. But our knowledge is second-hand, based on observing for ourselves the words those other observers wrote. (Or copies.) Reading-knowledge runs the gamut from learning what a friend saw a week ago in Paris by reading the letter she sent to us in New York, to knowing what Munn saw when he copied Skinner's words that scientists do not rely on words, to knowing what Einstein saw when he read the writings of Newton, Maxwell, Mach, Planck, etc. Our second-hand knowledge of what others observe is based on first-hand interpretation of what we ourselves have observed . . . which is what?

Digging into the most basic of all our basic beliefs is essential for refining our everyday thinking. Realizing that we do not get our concepts of things by sensing them forces us to look elsewhere for their source. Rightly or wrongly, Einstein insisted that we ourselves are the source.

Theory-laden observation: Einstein's two components

Those who have studied 'scientific observation' have coined a phrase for the kind of second-thinking needed to understand that scientists do not literally see what they naively think they see. The phrase is "theory-laden observation" or "theory-laden perception." But 'it' is really two distinct acts, the same two that Einstein referred to, namely, sense experience distinct from the simultaneous conceptual thinking or interpretation.

Their idea is basic common sense. What old-enough child cannot understand "See Santa over there? That's not really Santa." What old-enough student of the sciences cannot understand "The next time you watch the sun rise, think about the fact that the sun isn't really rising." It is the same common sense used to enjoy a long-ago Peanuts cartoon strip by Charles Schultz where Lucy, Linus, and Charlie Brown are stretched out on a hillock, gazing up at the passing clouds. Lucy comments, "If you use your imagination, you can see lots of things in the cloud formations." Linus 'goes first.' "Well, those clouds up there look to me like the map of the British Honduras in the Caribbean . . . that cloud up there looks a little like the profile of Thomas Eakins, the famous painter and sculptor . . . and that group of clouds over there gives me the impression of the stoning of Stephen . . . I can see the Apostle Paul standing there to one side." Asked what he sees, Charlie Brown replies, "Well, I was going to say I saw a ducky and a horsie, but I changed my mind."

But what they saw were clouds. The clouds did not become something else when they were looked at. Every one of the ideas, from the map to the horsie, came from Linus's and Charlie's two private stocks of old ideas stored in memory. That is science as good as anything connected with nuclear research.

> Physicists often speak of "seeing" the subatomic particles that are produced in particle accelerators. The particles can be made to interact within a device called a bubble chamber, which is a large tank that has been filled with liquid hydrogen. As the particles pass through the hydrogen, they produce tiny bubbles. After these bubbles are photographed, the particle interactions that took place can be studied at leisure. But are the particles really seen? Of course not. The only things that are observed are hydrogen bubbles that are strung together like beads. (R. Morris, *Dismantling the Universe*, p.203)

Even that, as we will see, is only part way to the full answer. Photos of hydrogen bubbles are not hydrogen bubbles. But how could readers learn what physicists see without the ideas 'behind' such words as see, full answer, and photo?

Reading: The Scientific-Method Paradigm

Pragmatic fictions: useful, even indispensable concepts

Applying Ockham's Razor, therefore, must not be taken too far. We create extra concepts because we need them. For instance, imagine yourself being given two balls and told to throw them as far as you can. Imagine one is a genuine golf ball and the other is a hollow, plastic, fake golf ball. If, as physicists do, you try to explain the two different paths they take when they leave your hand, you will need at least the following concepts: size, shape, weight, distance, decreasing height, inertia, gravity, and friction. The index of any college textbook on physics will have hundreds of names for concepts deemed useful for explaining things in the physical world that no one has ever observed.

The same is true for anyone who wishes to explain, not the physical things we try to learn about, but the concepts and theories we create as we learn about them. The concepts and theories are as invisible and intangible as the atoms or subatomic particles, and concepts must be invented to learn about them. Among the ideas needed to think about invisible ideas are those named by such non-words as sentence, subject, verb, predicate, direct object, indirect object, adjective, adverb, and so on. The ideas create, imagination, idea, concept, hypothesis, and so on are also needed.

Are those language fictions, psychological fictions, or just plain fictions?

E. CONCLUSION

Not easy to take seriously, even harder to believe

If all 'higher learning' observation is theory-laden, then the next step must be to become super-conscious of the unnoticed but powerful roles that radically diverse theories play, especially when different people are asked "What do you see?"

Einstein saw deeply into the role played by long-established habit. The only reason anyone believes — falsely — that they see numbers, words, the moon, hands, etc., is their long-standing thought-habits. Those

68

habits are more than just 'not-easily-noticed.' Entertaining the possibility that those thoughts are not true is very hard. Examining all of the evidence open-mindedly and becoming convinced that those long-standing, common-sense thoughts are literally false is next to impossible for people too far along in their learning careers to start over.

Specifically, to look at our hand (that is at what we mis-take for our hand) and try to believe, "That's not a hand" is hard. To do the same with respect to the moon is just as hard. That is why beginning with numbers and words is easier, and offers an opportunity to practice noticing 'the gulf' and understanding why great thinkers from Descartes to Einstein agreed that we have no sense contact with the physical world.

But, if we can learn to 'see' what we really see while watching a movie (those are just patterned colors on the screen) and while looking at 5 and same, we can also acquire the more general, true-thought habit. Others have done it before us. There is no reason we cannot do what they did, however difficult it is at first.

CHAPTER III-III.

Reading: Mentalities/Mindsets

Furthermore, after some vain efforts, I discovered
that the mentality which underlies a few of the essays
differs so radically from my own, that I am incapable
of saying anything useful about them.

A. Einstein, "Reply to Criticisms."

A. MENTALITIES, WORLDVIEWS, SYSTEMS

Question: What are the meanings of what is written?

Common-sensically speaking, it seems clear that there are thousands
of different languages, i.e., endlessly varied sounds and colored figures.
This has not prevented their users from understanding one another. We in
the twentieth–first century can use library resources to discover for
ourselves what Plato and Aristotle were thinking centuries ago when they
'wrote Greek,' what Augustine and Aquinas were thinking years ago
when they 'wrote Latin,' what Descartes was thinking years ago when he
was 'writing French,' and what Kant and Einstein were thinking when
they were 'writing German.'

Once Descartes concluded that recent discoveries prove that physical
things cannot be sensed, thinkers — most notoriously George Berkeley
— began to create radically new belief systems. Because none of them
could create new systems except by building on their original common-
sense philosophy, the new system-builders had to use 'common-sense
language' to express their new ideas. "Apple," to take just one example,
might now be used for a physical collection of colorless, odorless,
tasteless atoms. Or it might refer to an immaterial collection of color,

odor, and taste sensations. It might even be used for an aspect of cosmic thought.

The 'Linguistic Turn'

Since ordinary words could no longer be relied on to have just one stable meaning, armies of twentieth–century thinkers sought for a way to 'scientifically' determine words' meaning. But they failed to realize — really realize! — that words and language as such do not exist. That is why, in contrast to older thinkers who focused on thought, the new students of language turned attention from thought to language. For mentally grouping these widespread efforts, G. Bergmann coined the two-word phrase, "Linguistic Turn." Valuable lessons can be learned from their efforts.

Plan

Based on the theories of the 'language' philosophers, this third and final section of Chapter III will present three corollaries to the chapter's earlier segments. All three assume that most higher or scientific learning occurs during the process of reading. Not, by any means, all of it, just most of it. (Consider the way things are learned via the internet.)

B. COROLLARY ONE: EVERYONE CAN UNDERSTAND EVERYONE ELSE

Logical Positivist 'philosophers'

Many of those who turned to an intense study of language's meaning were the thinkers who regarded themselves as philosophers. To understand some of the startling hypotheses they created, an introductory mini-course is mandatory. It will center on only one sub-set of the 'linguistic turners,' the logical positivists.

The thesis of the logical positivists was revolutionary in the extreme. According to the most radical of them, the writings of Plato and the most famous thinkers of the past were not words with meanings. They were ciphers without meaning. What those old 'metaphysicians' wrote seemed to make sense, but only because what they wrote looked like words that really do make sense. Take 'Twas brillig, and the slithy toves did gyre and gimble in the wabe," for instance. It looks and is pronounced like things that make sense. If we heard it often enough, and if its separate components were woven into the fabric of everyday talk, we might think it actually does have a meaning. But it doesn't. Nor do Plato's non-words.

The most famous claim by a member of the group was A. J. Ayer's assertion that "God does not exist" makes no more sense than "God does exist." He held that neither makes any more sense than "Bodies are not the same kind of things as souls," another allegedly meaningless sentence. Ayer's proof was the 'verifiability principle." According to it, the only statements about reality that were meaningful were statements that could be tested and verified.

Ayer (1910-89) was only one of a large number of thinkers who contributed to the Linguistic Turn. His *Language, Truth and Logic*, published in 1936, was a late-comer to the debates about language's meaning. Ideas that others often couched in dense, esoteric terminology and pseudo-mathematical formulas, Ayer stated clearly and boldly. His book triggered a flood of articles and books, both pro and con. In the end, the cons 'wore down' the pros.

That is, it was not long until the verifiability principle lost favor. It was found to be unverifiable, hence metaphysical, hence meaningless, and hence unable to be understood. In time, it was disguised as the falsifiability test, according to which a theory is meaningful and scientific, only if it can be tested and proven false.

Logical Positivism's metamorphosis

The verdict in this book is the following. The idea that verifiability and/or falsifiability should be used as tests to see whether a statement has a meaning or is scientific as opposed to metaphysical or religious fails on two counts. First, the ideas are neither verifiable nor falsifiable. They are just pieces of self-refutation, more esoteric than easier-to-understand examples, such as "This statement is false," "I can prove that I don't exist," or "An unjust law is no law at all."

Reading: Mentalities/Mindsets

The second reason such 'philosophical' theories fail is because they can be understood, even if they are neither verifiable nor falsifiable. It takes only a bit of explanation for everyday thinkers to understand the verifiability rule and the falsifiability criterion. The meanings of both are as intelligible as "There's no god," "There is a god," "Elves exist, but Santa Claus doesn't," etc.

Both theories were definitively rebutted by W. V. O. Quine (1908-2000). In a famous 1950 essay, "Two Dogmas of Empiricism," he used his familiarity with probably hundreds of theories proposed by hundreds of 'linguistic turners' to draw his conclusion that, not only are statements verifiable, they are all true!

> Any statement can be held true come what may, if we make drastic enough adjustments elsewhere in the system. Even a statement very close to the periphery can be held true in the face of recalcitrant experience by pleading hallucination or by amending certain statements of the kind called logical laws. Conversely, by the same token, no statement is immune to revision. (W. V. O. *From a Logical Point of View*, p.43)

By the 1980's, the debates about whose claims can be understood, and whose claims cannot be, would have been finished, had everyone agreed that Quine had definitively falsified the theories of the logical positivists by pointing out that they can be verified. For some, however, personal experience weighed more heavily than bad theory. A. J. Ayer, for instance, reportedly had an experience which shook his faith in theories about whose claims can be understood and whose cannot. His 6-29-89 obituary in the *New York Times* reported that report.

> Sir Alfred, an atheist, was so persuasive in argument, the story goes, that when the English writer Somerset Maugham lay dying, he got Sir Alfred to visit him and reassure him that there was no life after death.

> In 1988, Sir Alfred's heart stopped for four minutes at a hospital in London, and he wrote later that he had seen a red

light and become "aware that this light was responsible for the government of the universe."

The experience left his atheism unquenched, he wrote, but "slightly weakened my conviction that my genuine death — which is due fairly soon — will be the end of me, though I continue to hope it will be." (E. Pace, "A. J. Ayer Dead in Britain at 78; Philosopher of Logical Positivism.")

Those who remain convinced that there is no longer any "me" named Ayer, can be expected, almost guaranteed, to invoke Quine's reference to hallucination to protect their naturalism.

So far as the meaningfulness of language is concerned, however, only one principle works for every reader. It was stated by Richard Rorty nearly thirty years after Quine's essay. It may even have been aimed specifically at A. J. Ayer's youthful claim that atheism and theism are equally unintelligible.

One philosophical method which will do no good at all is "analysis of meanings." Everybody understands everybody else's meanings very well indeed. The problem is that one side thinks there are too many meanings around and the other side too few. In this respect the closest analogy one can find is the conflict between the inspired theists and the uninspired atheists. An inspired theist, let us say, is one who "just knows" that there are supernatural beings which play certain explanatory roles in accounting for natural phenomena. . . . The atheists view these theists as having too many words in their language and too many meanings to bother about. Enthusiastic atheists explain to inspired theists that "all there really is is . . . ," and the theists reply that one should realize that there are more things in heaven and earth. . . . And so it goes. The philosophers on both sides may analyze meanings until they are blue in the face . . . (R. Rorty, *Philosophy and the Mirror of Nature*, pp. 88-89)

Reading: Mentalities/Mindsets

"Everybody understands everybody else's meanings very well indeed." We can call this "Rorty's Rule." Or, we can call it the *Intelligo* (Latin for "I understand"). It can take its place beside Descartes' *Cogito* (Latin for "I think"), shorthand for his claim that the thought we can be most certain is true is "*Cogito, ergo sum*" (Latin for "I think, therefore I am"). Even those who do not believe the Cogito is true can understand it. "True" and "intelligible" are not synonyms.

There are still many logical positivists. Most have simply changed their name and the name of their theory to "naturalist" and "naturalism." Their ideas continue to damage many people's thinking. For instance, the dominant reason for claiming that science and scientific theories are radically distinct from religious teachings is that, supposedly, the former can be falsified, whereas the latter cannot.

C. COROLLARY TWO: THOUGHT IS MORE THAN 'PUBLIC' BEHAVIOR.

'Scientific' psychologists

Nothing is more common-sensical, everyday-thinking than the idea that our thoughts are private to us. Not only private, but invisible and intangible. Thinking, of course, is regarded by most of us as the fundamental activity of the human mind. What, then, can we learn from those expert students of the human mind, scientific psychologists?

Probably the most accessible, one-volume history of their theories is M. Hunt's 1993 *Story of Psychology*. He begins with a prologue describing the ideas of seventh century B. C. ancestors, proceeds with three chapters titled "Prescientific Psychology" (from the ancient Greeks to Kant), follows them with seven chapters on "Founders of a New Science" (from Mesmer to the Gestaltists), and subdivides the rest of the new-science psychologists into eight categories of narrow specializers. His 650-page history is a bare-bones survey of literally thousands of thinkers whose views are as utterly contradictory as those of the 'philosophers.'

The Wonderful Myth Called Science

Unfortunately, 'the public,' taught that philosophers disagree about everything, are not taught that the same is true of the 'psychologists.' 'The public' generally conceives psychology in the same mythical way that they conceive science generally, that is, as a collective body of proven knowledge. Hunt's own story is slightly marred by that myth. None of the pre-1800 thinkers he regards as pre-scientific ever regarded themselves as anything but scientific. They all included psychology as an important part of their respective worldviews. For instance, it was the great philosopher-psychologist Plato who invented the first scientific model for studying the psyche. (See Book IV of his *Republic*.) He also invented the distinction between what is sensed and how we conceptually interpret what we sense, the 'primary' distinction used by Einstein to build his theory about scientists' theories!

What actually (truly) happened in post-1800 was that those who had fallen in with the habit of reserving "science" for new theories achieved by the mythical 'scientific method,' began to demand a niche in the university distinct from that occupied by 'philosophers' who taught what is now called "philosophical psychology." The innovators succeeded, which meant that they were allowed to do what physicists, chemists, and biologists were allowed to do, namely, to proceed without bothering about the everyday questions dismissed as "metaphysical."

Looking-outward 'behaviorists'

Early 1800's 'scientific' psychologists took consciousness as their special subject-matter. To study sensation, emotion, and so on — everything usually grouped under "consciousness" — they cultivated a special method that they called "introspection." The only consciousness anyone has direct access to is one's own, and the only way to study it is to 'look into' one's own mind.

In the late 1800's, John Watson (1878-1958) read the reports of those introspectors, saw how deeply divided their conclusions were, and declared that scientific psychologists should start all over again. Watson said they should use the same method other scientists use. Astronomers, for instance, do not learn about the heavens by going into a closet to do intense navel-gazing. They use normal, outward-facing observation. That is, they look. What can outward-facing psychologists look at and observe? Watson's 1913 answer was "Behavior!"

Soon, Watson's behaviorist program did in psychology departments what the logical positivist program had done in philosophy departments, viz., gained countless converts.

> It is an interesting exercise to sit down and try to be conscious of what it means to say that consciousness does not exist. History has not recorded whether or not this feat was attempted by the early behaviorists. But it has recorded everywhere and in large the enormous influence which the doctrine that consciousness does not exist has had on psychology in this century. What a startling doctrine! But the really surprising thing is that, starting off almost as a flying whim, it grew into a movement that occupied center stage in psychology from about 1920 to 1960. (J. Jaynes, *The Origin of Consciousness in the Breakdown of the Bicameral Mind*, pp.13-14)

Naturally, the behaviorist program had a far-reaching, negative impact on the common-sense idea that our ideas and feelings are private. In order to smooth the way into their absurdly anti-common-sensical theory, Watson and others continued using the old vocabulary. They continued to talk about perception, anger, curiosity, love, even thinking. But they orchestrated a whole-sale renaming project. The old names were detached from their common-sense use for inside feelings, memory-images, and thoughts, and pasted onto various types of outward behaviors.

The most crucial of the re-definitions was for "thinking." Behaviorists re-defined "thinking" the way many of the language philosophers did, namely, as talking. The personal thinking of psychologists themselves became 'out of bounds.' Theories about thinking were replaced by theories about talk, and the latter were renamed "theories about thought." Psychologists determined to be scientific were, in effect, locked out of their own minds. References to private, subjective experience were censored, replaced by words for verbal behavior. We think out loud when we are speaking. When others are thinking/speaking out loud, we politely talk silently to ourselves. Thought, that is verbal behavior, takes two forms, overt and covert. But both forms are forms of behavior. As late as 1974, B. F. Skinner continued to insist that "The history of human thought is the history of what people have said and done." (See *About Behaviorism*, p.130)

The great 'Lock Out' began to weaken after the mid-1900's. In time, what is now referred to as "the Cognitive Revolution in Psychology" took place. A major contributor to the revolution against behaviorism was Noam Chomsky, most often identified as a linguist. His scathing review in 1959 of B. F. Skinner's *Verbal Behavior* had a far greater impact on psychology than Skinner's book itself had.

It should be better known than it is that there is utter chaos in the 'field' of scientific psychology. In 1988, many members of the *American Psychological Association* broke away and formed the *American Psychological Society*. According to a later article in the *New York Times* (March 9, 2004), Society members accused Association members of not being scientifically rigorous. The report added that the Association's president-elect described Society members as "being overly devoted to the scientific method"! Subsequent to last summer's meeting of the Association, the *New York Times* (August 10, 2004) ran another article entitled "For Psychotherapy's Claims, Skeptics Demand Proof." In the opinion of one observer, "The split in the field is bigger than it ever, has ever been."

The metamorphosis of behaviorism

As in the case of the logical positivists, most behaviorists did little but change their name. They now identify themselves as cognitive-behaviorists or as cognitive psychologists. As for introspection, some now admit that it is legitimate. But that word, too, has been re-defined. Objective researchers cannot base their conclusions on an inspection of their own subjective experience. They can, however, use others' 'introspective' verbal-reports (behavior) collected from tests administered to those others. Whereas behaviorists could think, or rather talk, only about others' verbal behavior itself, cognitivists are permitted to use others' verbal behavior as data for hypotheses about the test-taking others' unseen cognitions and/or consciousness.

B. F. Skinner was not deceived. Just three years before his death (1990), he reportedly told a *New York Times* interviewer, "I think cognitive psychology is a great hoax and a fraud, and that goes for brain science, too." (*New York Times*, 9/25/87) Little wonder that Gordon Allport wrote that psychology "progresses in fits and starts, largely under the spur of fashion."

What can psychologists contribute positively to the language-related question, "What is meaning?" The answer is "Very little." The greatest contribution to questions about meaning and thought, the real thing, came from William James, but he is more often dismissed as a philosopher, not hailed as a vitally important, genuinely scientific psychologist. Nor will matters change until the myth of science-in-general is renounced and the chaos in 'psychology' is seen for what it is, namely, a parallel to the chaos in 'philosophy.' The dream of creating a philosophy-free way to study people has proven only that there is no such thing. Sherlock, of course, would regard that as a valuable lesson.

It would be hard to find a more perfect conclusion for this snippet of history than a statement from a leader of the cognitive revolution, Ulric Neisser. He made this statement during an interview published in 1986:

> . . . the fact that we understand so little about reading after years of study is very impressive to me. Many people think we do understand it; there's an awful lot of theory about automatic processes, decoding, storing, and so on. It doesn't seem to me, though, that those models are particularly helpful...
>
> The contrast [of cognitivism] with behaviorism is still striking. It's great to be free to talk about thinking and memory, seeing, knowing, and imagining, without having to look over your shoulder every minute, without having to

explain that when you go back to your lab you won't do it any more. It's great not to have to use the vocabulary of implicit responses, not to have to pretend that all learning is due to reinforcement, or, alternately, to spend lifetimes proving that learning can occur without it. There were people who spent decades trying to prove that, as if they had never engaged in a conversation and remembered it afterward. It was a painful waste of talent and energy, imposed by behaviorism for 30 years or so. At least we're free of that, we don't have to do it anymore. (Ulric Neisser in B. Baars, *The Cognitive Revolution in Psychology*, pp.282-83.)

This book should be required reading for every student enrolling as a scientific-psychology major.

The only first-hand study of thought you, like any of us, can engage in is a study of your own private thinking. You *are* understanding thoughts as you scan these non-word 'words,' right?

D. COROLLARY THREE: MINDSET IS CONTEXT.

Every word-user has a unique mentality or mindset.

This third, pivotal chapter began with a quote from Einstein, saying that he had spent many happy hours reading the works of Bertrand Russell. Moreover, he was easily persuaded to write an essay for the 1944 volume, *The Philosophy of Bertrand Russell*. From what Pais, Einstein's biographer, reported, Einstein and Russell shared an interest in the nature of human knowledge:

He [Einstein] studied philosophical writings throughout his life, beginning in his high school days, when he first read Kant. In 1943 Einstein, Gödel, Bertrand Russell, and Pauli gathered at Einstein's home to discuss philosophy of science about half a dozen times. 'Science without epistemology is—in so far as it is thinkable at all —primitive and muddled,' he wrote in his later years . . . (A. Pais, *"Subtle is the Lord..." The*

Reading: Mentalities/Mindsets

Science and the Life of Albert Einstein, Oxford: Oxford U. P., 1982, p.13.)

We can say, therefore, that Einstein and Russell had similar mentalities. By contrast, some people's worldviews were so radically different from Einstein's that he declined to comment on them. Referring to some of the essays contributed to the Schilpp volume entitled *Albert Einstein: Philosopher-Scientist,* he wrote:

> Furthermore, after some vain efforts, I discovered that the mentality which underlies a few of the essays differs so radically from my own, that I am incapable of saying anything useful about them. (A. Einstein, "Reply to Criticisms.")

The reason for introducing these references to 'mentalities' is to point out a major factor that influences everyone who reads. Quite simply, it is impossible for anyone to read with a blank mind. For 'words' to evoke any response from a reader who scans them, the reader's mind must already be primed with an enormous amount of knowledge. That knowledge will have at its core a common-sense philosophy or worldview similar to everyone else's. Constructed around it will be volumes of personal memory, as well as those 'primary' conceptual features that serve as a framework for the reader's mature belief-system. Call that mindset or personal worldview of the reader "his or her mentality or mindset."

When people's mindsets differ radically enough, it is almost as if they 'see' different things, whether they are looking at the night sky or scanning words in a book. When those readers later write for other readers, the non-word 'words' they use can have unusual meanings. One way to put this thought 'into words' is to say that mindset is context for word-users. The context for readers is their unique mentalities. The context for writers is theirs. When the two contexts diverge sufficiently, 'communication' becomes difficult, unprofitable, or impossible.

The Wonderful Myth Called Science

Kuhn's provocative claim

The more strong-habited are the components of a reader's belief-system, the more those habits will impact the reader's perception of what he or she reads. One of Thomas Kuhn's more provocative views was his suggestion that, when our distant ancestors looked into the night sky, they 'saw' different things from what we 'see.' Where they 'saw' only stars, we 'see' planets and galaxies as well as stars. In the 1970 postscript to *The Structure of Scientific Revolutions*, Kuhn explains what he (probably?) had in mind when he made his claim.

> If two people stand at the same place and gaze in the same direction, we must, under pain of solipsism, conclude that they receive similar stimuli. . . . But people do not see stimuli; . . . the route from stimuli to sensation is in part conditioned by education. (T. Kuhn, *The Structure of Scientific Revolutions*, p.192-93)

A major aim of this third chapter is to invite its readers to apply a parallel thesis to the 'seeing' done by readers of texts. Simply replace Kuhn's "sensation" with "theory-laden perception."

The mindset-seeking 'literary critics'

One picture is worth a thousand words. That thought is true, but only if the person who sees the picture has a mindset primed to 'see' what the picture is meant to represent.

Here, then, is a picture to introduce a major thesis of this book. Imagine yourself standing in the middle of a library which holds copies of all the books, journals, and newspapers every written. Behind every book and every journal or newspaper article is a human author with his or her unique mindset. No human being could possibly learn and then remember every complete thought that ever went through any of those thousands of authors' minds as they penned complete sentence after complete sentence.

But imagine that one of us humans does try to do just that, viz., to look around and to 'see' in a glance all the aspects of all the thoughts found in all of the library's texts. Save that picture. It will help you appreciate what another group of language students set out to do. They call their enterprise "literary criticism." They read texts by the hundreds or thousands. They dig out the mindsets of the writers. They classify, i.e., mentally group them. They write books to explain their findings. And then . . .

And then they read each others' books. They study other critics' mindsets. They classify, i.e., mentally group them. They write books to introduce groups of students (classes) to the enterprise itself, viz., to literary criticism.

The result: Anthologies designed to introduce course-taking students to what might be found 'in' or 'behind' the library's texts. For example, someone named "Shakespeare" (heaps of texts just on that!) wrote a play, King Lear, about imaginary people who never lived. (I have a text whose author says so.) A literary-critic-anthology writer will notice that some of the other literary critics have a Marxist worldview and are especially inclined to interpret Shakespeare's text from that personal Marxist standpoint and especially inclined to interpret Lear-readers who write texts commenting on the Lear text from that personal Marxist standpoint. (To keep matters simple, temporarily ignore the worldviews of readers who do not write texts.) Other literary critics are found to have a worldview shaped by Freud's theories which leads to their theories about the author's mindset and those of the reader-commentators on the Lear text. Still others have a feminist worldview, or a structuralist worldview, or a post-structuralist one, and so on.

For instance, D. E. Hall's 2001 *Literary and Cultural Theory* is a recent introduction to the 'field.' It reflects the extraordinarily wide-range of Hall's text-reading, as well as a classification system helpful to someone hitherto unfamiliar with the vast, textual output by literary critics. After a few pages introducing the "long, dense, and complex" history of human writers' literary creations and human readers' responses to those creations, Hall divides into separate mental groups the theories produced by those who have studied those creations and responses. The titles of his first three chapters will give some sense of what is meant.

The Wonderful Myth Called Science

They are "The New Criticism and Formalistic Analysis," "Reader-Response Analysis," and "Marxist and Materialist Analysis."

Critical readers must emulate literary critics

Literary critics now constitute the largest crowd of experts who concern themselves chiefly with texts or what is written. They are all individuals, and all have individual theories which serve as the lens used for their theory-laden reading of texts. If mentalities or mindsets are crucial factors in every instance of word-use, it may be possible for us to learn from them by becoming super-conscious of the mindsets 'behind' every text we ever read, whether it purports to be written by a scientist, a philosopher, a theologian, a literary critic, etc. Especially because *every text we read comes from a reader.*

And every one of those readers who has written a text has or had a unique mindset, shaped by a lifetime of personal experience and learning.

Toward a generalizing principle

Interestingly, a generalizing principle enunciated by the medieval thinkers can be of assistance here. Those not-so-long-ago thinkers wrote, *"Quidquid recipitur, per modum recipientis recipitur,"* Latin for "Whatever is received is received after the manner of the receiver." A Christian who sees "Crusades" responds very differently from a Muslim who sees it. As always, the determining factor is nothing less than a part of the receiving or perceiving reader's worldview, overall belief-system, philosophy, etc. —the name isn't critical, the thought is.

No passage more perfectly captures the medievals' *"Quidquid recipitur..."* than the opening paragraphs of Alan Watts's book, *Nature, Man and Woman.*

> A floor of many-colored pebbles lies beneath clear water, with fish at first noticed only by their shadows, hanging motionless or flashing through the liquid, ever-changing net of sunlight. We can watch it for hours, taken clear out of time and

85

our own urgent history, by a scene which has been going on just like this for perhaps two million years. At times, it catches us right below the heart with an ache of nostalgia and delight compounded, when it seems that this is, after all, the world of sane, enduring reality from which we are somehow in exile.

But the feeling does not last because we know better. We know that the fish swim in constant fear of their lives, that they hang motionless so as not to be seen, and dart into motion because they are just nerves, startled into a jump by the tiniest ghost of an alarm. We know that the "love of nature" is a sentimental fascination with surfaces—that the gulls do not float in the sky for delight but in watchful hunger for fish, that the golden bees do not dream in the lilies but call as routinely for honey as collection agents for rent, and that the squirrels romping, as it seems, freely and joyously through the branches, are just frustrated little balls of appetite and fear. We know that the peaceful rationality, the relaxed culture, and easy normality of civilized human life are a crust of habit repressing emotions too violent or poignant for most of us to stand—the first resting place which life has found in its arduous climb from the primordial, natural world of relentless struggle and terror.

But we think we know. . . Our view of nature is largely a matter of changing intellectual and literary fashion. . . (A. Watts, *Nature, Man and Woman*, 1952)

Almost, but not quite. There is no 'our' view of nature. There are only individuals' views.

Tear down those mythical fences.

For twenty–first century thinkers whose goal is a unified and true worldview, an enormous intellectual change is required. Every library's holdings are divided according to some individual(s)' arbitrary though not irrational grouping system. The Library of Congress classification system begins with the A's for general works, the B's for philosophy and religion, the C's for auxiliary sciences of history, and ends with the U's for military science, the V's for naval science, and the Z's for bibliography. That system replaced the Dewey Decimal system which ran from the 000's to the 900's. Those are different arbitrary groupings.

Corresponding somewhat to those arbitrary groupings, all of us have been 'brainwashed' into believing there are autonomous bodies of knowledge named science, philosophy, theology (religion), common sense, etc. These are arbitrary, no-longer-useful, progress-impeding, mental groupings that hang around the collective neck of academia like a huge, stupid albatross. Not till the truth about human knowledges — individual and plural — is learned by every learner will the dead albatross fall away.

Literary Theory: An Introduction, by Terry Eagleton, offers a corrective. After an Introduction entitled "What is Literature?" Eagleton groups his targeted subjects or texts or themes into five classes, named "The Rise of English" (when did it become a separate college-and-university discipline?), "Phenomenology, Hermeneutics, Reception Theory," "Structuralism and Semiotics," "Post-Structuralism," and "Psychoanalysis." Standard fare perhaps, but wait, Eagleton's text brings out the more general idea of theory-laden perception.

> What we have uncovered so far, then, is not only that literature does not exist in the sense that insects do, and that the value-judgments by which it is constituted are historically variable, but that these value-judgments themselves have a close relation to social ideologies. (T. Eagleton, *Literary Theory*, p.16)

Even better, Eagleton's text exposes the transparently mythical lines between theories.

> The economist J. M. Keynes once remarked that those economists who disliked theory, or claimed to get along better without it, were simply in the grip of an older theory. This is also true of literary students and critics. There are some who complain that literary theory is impossibly esoteric — who

suspect it as an arcane, elitist enclave somewhat akin to nuclear physics. It is true that a 'literary education' does not exactly encourage analytical thought; but literary theory is in fact no more difficult than many theoretical enquiries, and a good deal easier than some. I hope this book may help to demystify those who fear the subject is beyond their reach. Some students and critics also protest that literary theory 'gets in between the reader and the work.' The simple response to this is that without some kind of theory, however unreflective and implicit, we would not know what a 'literary work' was in the first place, or how we were to read it. Hostility to theory usually means an opposition to other people's theories and oblivion of one's own. One purpose of this book is to lift that repression and allow us to remember. (T. Eagleton, *Literary Theory*, pp. vii-viii)

Physicists, chemists, biologists, psychologists, literary critics, and first-semester 'philosophy' students may not think they have a philosophy. That opinion, or rather that illusion, is part of their philosophy or worldview. One they had not yet acquired the day they were born. And one that certainly 'gets in the way' of what they 'see,' as Kuhn, like Einstein, emphasized. Alasdair MacIntyre put it nicely:

Philosophy just is conceptually self-conscious enquiry in whatever field. There are philosophical physicists, historians, linguists, theologians, and psychologists; but 'the' philosopher, who is philosophical *an sich*, but not any of these, nor a philosophical mathematician nor a . . . (the list is as long and as indeterminate as are the descriptions of intellectual enquiry) is a mythological beast. (A.MacIntyre, "Philosophy, the 'Other' Discipline." *Soundings*, v.65, 127)

What do readers observe, then? If, in Eagleton's terms, literature does not exist the way insects do, do texts? Is "text" synonymous with "book"? Can there be books, if there are no words? If "language" is the name for the last of the great myths, how can there be either texts or books? Can we learn from myths?

The Wonderful Myth Called Science

E. SUMMARIZATION

Everyday common-sense is the foundation for all 'higher' learning.

We cannot begin 'formal education' until we have acquired an inner model of outer reality and begun to associate non-word 'names' with the items in that inner model. Whenever you take in hand a text whose author believes in any other foundation than that early common-sense worldview which every normal learner has by the age of five or six, be on your guard and especially critical.

That foundation — common- sense — is thoughts or meanings.

Common sense is theory. Five-year and six-year olds have a philosophy. It is a huge network of beliefs, both tacit and explicit, which forms a platform for later specialized learning. Keynes's remark about the economists who did not notice that they were in the grip of an older theory applies here. Those who regard common sense as pre-conceptual are blind to the obvious fact that it is highly-complex conceptual theory.

Equally mistaken is the popular theory that all theories and beliefs are 'socially constructed' or 'cultural.' Stripped of their local-cultural beliefs, five-year and six-year olds would not thereby become belief-less. Exactly the reverse is true. Without the complex theories she or he acquires by an early age, no human learner could acquire any specific-cultural beliefs. In fact, the belief that all beliefs are specifically cultural — or confined to a specific historical era or 'true' only relative to a specific place — is itself a specific cultural belief. Whoever unwittingly becomes a victim to that myth can learn from the example of Socrates who rejected the similar half-truth preached as the whole truth by his and Plato's greatest adversaries, the Sophists.

Common-sense beliefs contradict each other.

The fact that common-sensical everyday thinking is the foundation for all later theorizing, beginning with that of the physicists, does not mean that all of it is the truth about the world. The truth is just the

89

reverse. It is precisely the 'internal contradictions' in our everyday thinking that open the door to the discovery of common-sense illusions. Not every understood-thought is a true thought.

In his *Meditations*, Descartes asked the same kind of question asked by the West's first scientist-philosopher, Thales: What really exists? Like Thales, Descartes concluded that what we sense-experience are appearances of what does not appear. Descartes' sample was what appears to be wax. When melted, it changes in every sensible respect. Does the original wax still exist and, if so, what is it? Descartes could never have asked that question, if he had not already possessed everyday-thinking ideas of wax and wax melting. Nor could he have asked whether the sun is not indescribably larger than the moon rather than apparently as small, if he did not already have ideas of the sun, the moon, and distance. Nor could he have wondered whether it is possible to still feel an amputated arm that no longer exists, had he no prior understanding of such concepts as size, arms, armless stumps, pain, etc.

But, precisely because Descartes — the most revolutionary thinker of all time and, as a result, the initiator of 'modern philosophy' — begins with common-sense, many readers object that his reasoning refutes itself. He is compared to the tree-climber who saws off the branch on which the tree-climber is perched. According to these common-sense readers, using the everyday-thinking belief that we hear thunder at varying times after we see the lightning to prove that we do not hear the thunder, only the sound waves that take time to reach our ears, seems to be destroying the common-sense premise from which the anti-common-sense conclusion is drawn.

However, immersion in reader-response theories proposed by medieval thinkers, literary critics, and Alan Watts prepares us to probe deeper to find the source of different readers' reactions.

The common-sense objection to Descartes is based on another common-sense assumption, namely, that there is no way to acquire ideas of physical things except by experiencing them. Einstein's insistence that we do not acquire our ideas of physical things by 'induction from sense experience' is the reason for his insistence that we create all of our concepts. The common-sense fact that we all have concepts of things we have never experienced as such — concepts that range from Santa Claus and Captain Picard to 'language' fictions — is a common-sense rebuttal of the common-sense assumption used against Descartes.

Our final G. U. T. will have to accommodate common-sense.

Until hopeless senility sets in, we will retain and largely rely on our original common-sense philosophy for everyday living. This will be the case, even if our final unifying philosophy must be partly anti-common-sensical.

This is not as hard as it might seem at first. As noted earlier, anyone can understand "Do you see Santa Claus over there? That's not Santa Claus, it's Uncle Harry." Philosophers at philosophy conventions can present papers in which one claims that there are only physical realities and the other claims there are no physical realities; both afterwards understand perfectly what is meant by "Would you like to go for a cup of coffee and discuss further whether coffee really exists?" And everyone with a five-year-old's worldview can understand the anti-common-sense pretending required to enjoy contrary-to-fact fairy tales.

Accommodating rival mindsets is accomplished by mindset-switching. When Einstein set out to explain his anti-common-sense relativity theories, he began by invoking contrary-to-fact pretending that relied on everyday thinking.

> We imagine a large portion of empty space, so far removed from stars and other appreciable masses that we have before us approximately the conditions required by the fundamental law of Galileo. It is then possible to choose a Galileian reference body for this part of space (world), relative to which points at rest remain at rest and points in motion continue permanently in uniform rectilinear motion. As reference body let us imagine a spacious chest resembling a room with an observer inside who is equipped with apparatus. Gravitation naturally does not exist for this observer. He must

91

fasten himself with strings to the floor; otherwise the slightest impact against the floor will cause him to rise slowly towards the ceiling of the room. (A. Einstein, *Relativity: the Special and General Theory*, ch. XX)

The student who tries to gain access to Einstein's anti-common-sensical mindset can do so only by making use of the everyday-thinking habits she or he brings into the classroom, and of pretending. The student may gain understanding from their habitual thoughts of the rising and setting of the sun, of the fall of apples and leaves from trees, of lightning bolts striking points on the ground, of resting elevators, sudden upward motion of elevators, pressure at the soles of one's feet and so on. The answer to "Why can we take physics courses without first taking a philosophy course?" is "Because we already have a common-sense philosophy." No student in any classroom begins with a philosophy-free mindset. And, of course, no reader of any book begins with a philosophy-free mindset, either.

So, whatever your current belief-habits are, understand this. The thoughts that will come to you as you read on are not for the faint of heart.

CHAPTER IV

Quintalism & Subatoms

To him who is a discoverer in this field, the products of his imagination appear so necessary and natural that he regards them, and would like to have them regarded by others, not as creations of thought but as given realities.

A. Einstein, *Ideas and Opinions*, p.264

A. NO MORE TIME FOR TWO-SIDED DEBATES

Space and time are not infinite.

That statement that space and time are not infinite can be easily misunderstood, especially by physicists. Therefore, let me explain what I mean by it.

There is not nearly enough space in a short book to deal in detail with every disputed question about 'science'. This is first thing that "space and time are not infinite" means in the present context. In addition consider the fact that space and time do not exist, and therefore they cannot be infinite. We do have concepts of both, in the same way that we have a concept of Santa Claus. But having an idea of something does not guarantee that it exists!

More challenging is the fact that there are literally no such things as concepts. I was twenty–nine when I began to suspect that the myth of concepts is the major obstacle on the road to genuine science, that is, on the road to having a belief-system whose 1001 basic convictions, circularly-argued to be logically consistent, are true.

But, even if concepts do not exist, complete thoughts do. We don't get concepts from seeing single words all by themselves. If I wrote science, knowledge, individual, ongoing, truth, belief, error, and so on, it would fit the logical positivists' criterion of non-sense. Right? Rather, books are made up of complete sentences, and the complete sentences are symbols for complete thoughts, not just single, isolated, detached concepts. In fact, this paragraph alone not only tells you what is true. It also gives you examples to prove it. Right?

What about "Right?" You see two items in that paragraph that look exactly like that third one which is part of this paragraph. Is "Right?" a complete sentence? Is it a single word? Is it five letters enclosed with quotation marks? Or are those really just eight more, similar sets of differently-shaped ink marks? Sets? So many confusing questions.

That is the kind of exercise that drives people insane. An interesting book, D. F. Wallace's *Everything and More* (2003) recently arrived. Wallace describes a case of what he calls "abstract thinking." In a manner reminiscent of Hume's discussion of induction, he invites us to picture ourselves in bed, thinking long and hard about why we feel confident that the floor will hold us up the next time we get out of bed and step on it. Just because it's done it a thousand times before, is no guarantee it will do it again. After a page of similar musing, Wallace concludes:

> Another sure sign it's abstract thinking: You haven't moved yet. It feels like tremendous energy and effort is being expended and you're still lying perfectly still. All this is going on in your mind. It's extremely weird; no wonder most people don't like it. It suddenly makes sense why the insane are so often represented as grabbing their head or beating it against something. If you had the right classes in school, however . . .
> (D. F. Wallace, *Everything and More*, p.14)

The Wonderful Myth Called Science

Such exercises won't drive you insane. But the possibility that much of our most fundamental thinking is superficial or even mistaken can be unsettling. In a revealing debate between himself and C. Rogers in 1962, Skinner — whose views were radically anti-common-sensical — referred to the curious behaviors exhibited by his students:

> I think you can make the assumption that each person is completely determined to do what he is now doing and is going to do by his own genetic and environmental history. I say this to my class at the end of the term. I give a course in human behavior with a fairly large number of students, and many of them become very anxious about this. In fact just this year I have learned two strange things about my course at Harvard: one that it sends more students to courses in the divinity school than any other undergraduate course; and also that it sends more to the health service. There is a recognizable syndrome that turns up. (In *Carl Rogers*: *Dialogues,* ed., H. Kirschenbaum & V. L. Henderson, pp.93-94)

In the pages that follow, it will be argued that, like governments, language, and solid flesh-colored hands, nearly all of the things that we grow up believing in, plus nearly all of the things that those in 'the sciences' believe in, are mythical. This is not only hard for us to believe —Einstein observed that it is equally difficult for theoretical physicists:

> To him who is a discoverer in this field, the products of his imagination appear so necessary and natural that he regards them, and would like to have them regarded by others, not as creations of thought but as given realities. (A. Einstein, *Ideas and Opinions*, p.264)

That is why what follows is not for the faint of heart.

Let me return to the real point of the opening reference to the brevity of space and time. In order to cover a large number of truly fundamental issues in a very brief book, a massive amount of material must necessarily be omitted. Half of what is omitted is important for showing the way all of the book's major conclusions grow out of convictions that are part of our original common-sense worldview. The other half of what is omitted is important for answering the classical objections to the theories of Descartes, Hume, Kant, and Einstein.

Let me make the last thought even clearer. The unifying framework proposed here will rely heavily on Ockham's Razor or Einstein's Simplicity. When any of us sees one of our favorite, most deeply-held convictions dismissed, particularly when it is dismissed without any evidence that it was ever taken seriously, we are likely to summarily dismiss the dogmatically proposed alternative. But, as a famous lecturer once remarked, "one can only make one point in one lecture," and it would require 1001 chapters to state and adequately defend just the most basic or 'primary' points of the unifying view argued for here.

This present book's unifying theory of what exists.

What exists? That is the bottom line for all of us. To unify modern thought, from Descartes to Einstein and beyond, we must learn to pretend that "What exists?" can be broken down into two questions: "What seems to exist?" and "What really exists?"

First, what seems to exist? Till the end of our sojourn here, the 'world' will continue to appear just the way it did when we were five or six. We will always be able to go outdoors on a clear night, look into the velvety black of distant space, and be once again awestruck by those tiny pinpoints of brightness so far, far away. If we stay up till dawn, we will see the sky brighten, see the reddish-golden sun come out of its tent and, as it rises to its zenith overhead, see it turn so blindingly white that it pains our eyes to stare at it. The daily bread of our meals will taste as we are used to it tasting. Drink of different kinds will slake our thirst as we expect from past experience it will. Scratching a new itch will afford a new pleasure, not a searing pain, just as boiling water spilled on exposed skin will produce pain, not pleasure.

But what really exists? The answer arrived at by the author of this book is best summarized this way. There are only five types of things in the entire universe — types, not individual entity-things. They are as follows.

The Wonderful Myth Called Science

Persons are the *most fundamental existents*. Their personal, private streams of consciousness, though unified in a way impossible to picture adequately, consist of three concurrent flows, a flow of sense-data, of memory-images 'of' previous sense-data, and ongoing complete thoughts. None of what we experience is physical at all. Those four types of things most certainly exist. If physical things exist at all, none of those for which we have only indirect empirical evidence are larger than subatomic particles.

In somewhat more detail, the five types of things are the following. What Einstein refers to in a variety of ways, very often as 'sensory experiences,' are effects in us. The most important are a wide field — a TVF — of patterned colors, seemingly wrap-around sounds, odors, tastes, feelings of warmth, cold, hardness, softness, roughness, smoothness, etc. Together, they are referred to as (i) "sense-data."[*] They are the first and most noticeable things we experience. In addition, there are four other types of realities. (ii) Image-copies of sense-data come in all five varieties that correspond to the five varieties of sense-data just referred to. (iii) The complete thoughts going through our mind are what constitute our own personal wisdom or folly.[†] (iv) We, the persons who experience sense-data, who are aware of images, and who understand thoughts, are the fourth category of things that really exist. Finally, if our ideas of physical bodies external-to and independent-of us are true in any

---

[*] There are various sense-datum theories. Here, the visual sense-datum is a total visual field or TVF. Imagine a driver who could see only one car ahead and not the road, or a basketball player who could see only the ball, but not the floor, teammates, etc.!

[†] The best description of ongoing complete thoughts in any of the world's literature is found in Chapters IX and X of William James 1890 *Principles of Psychology*

way, those physical bodies are all (v) subatomic particles capable of only two things, resting and moving relative to other things. I will call them "subatoms."

In order of importance, persons, their complete thoughts, images, and sense-data are most important. Even if subatoms exist, they provide nothing but a side-show. This reverses Thomas Huxley's thesis that consciousness is the useless side-show, produced as a powerless epiphenomenon or by-product of matter.

Like the views of other recent thinkers, this quintalist system is extremely reductionistic when compared with our everyday thinking. But it is far less reductionistic than Descartes' dualism and the opposing forms of monism, e.g., materialist-naturalism and idealist pantheism.

Common sense: experienced appearances versus underlying reality

Constant analysis, or at least a readiness to analyze, is required for all of those who have adjusted their final thinking to take account of Descartes' revolutionary discovery that naïve realism is internally contradictory.

Descartes himself points out the need for linguistic analysis. In Meditation Two, he incorporates the medieval distinction between things and their apparent qualities. We look down from our balcony and say "I see someone going by right now," even if the only thing we see is the umbrella that hides them from our gaze. He uses that to pave the way to realizing that the stiff, opaque, fragrant, cold piece of beeswax is the same 'thing' as the fluid, transparent, odorless, warm liquid in the heated pan. Beneath the surface qualities or sensible features our intellect 'discerns' the same 'whatever' that we began with. That 'whatever' is something wholly inaccessible to our senses. No one, Descartes was arguing, has ever seen wax as such.

Thales was the first thinker to introduce this scientific approach to reality. He did exactly the same that Descartes did, namely, recognized that slashing rain, softly-falling snow, rock-hard ice, oppressive humidity, dripping condensation, and rolling ocean waves are, despite the different appearances, one and the same thing. Every schoolgirl who studies chemistry and discovers the true nature of H_2O learns the same fundamental lesson.

Taking Thales seriously, however, means that not just ice, steam, and water are water, but that rocks, bones, and gristle are water, too. Taking Descartes seriously meant that air, earth, fire, and water are really different names for a single kind of thing, homogeneous granules of extended matter. Taking Rutherford's model of the atom seriously means that there are no rocks, no bones, and no air, earth, fire, or water, only tiny specks zipping about in nearly empty space. Taking Ockham's Shaving, Einstein's Simplifying, and Sherlock's Eliminating seriously means occasionally denying with words the existence of things we stop assenting to in thought. That is how the theory guiding this unification incorporates Einstein's advice:

> The whole of science is nothing more than a refinement of everyday thinking. It is for this reason that the critical thinking of the physicist cannot possibly be restricted to the examination of the concepts of his own specific field. He cannot proceed without considering critically a much more difficult problem, the problem of analyzing the nature of everyday thinking. (A. Einstein, *Opinions and Ideas*, p.283)

Translated here, his rule means that a critical examination of the concepts of one's own field must begin by showing that those concepts are solidly anchored in everyday thinking, even if, in the end, they require a rejection of certain elements of that original common-sense philosophy. Our final theory about what exists must be anchored, not in the full truth of naïve realism or common sense, but in the full meaning of everyday talk. That meaning is the ongoing everyday thought — the total ensemble of tacit and explicit beliefs — that we all understand so readily.

The rest of this book: one-sided unification from many debates

There will be no compromises from this point on. Each of us who discovers that we have no beliefs which have not been contradicted by many brilliant thinkers, must face up to our own personal responsibility. No one can make our decisions about "Who is right in this debate?" for us.

Mine have been made over the course of a rather long lifetime. At first, I acquiesced entirely in what I believed, from my reading, were Saint Thomas Aquinas's beliefs. So far as my ideas about how we acquire ideas were concerned, I was absolutely certain that the Latin adage, "*Nihil in intellectu nisi prius in sensu*" or "There is nothing in the intellect which was not first in the senses," was true. By the age of twenty–nine, I had finished constructing my tri-partite — science, philosophy, theology — worldview. It was only at that point that I discovered what Descartes discovered and what Einstein so often repeated, "None of our ideas are acquired or derived from what we sense." Once that foundation of my theory of knowledge was pulled from under my feet, I began the task of reconstruction. The modicum of brevity required here, in order to present the now-rebuilt worldview, brings with it a risk — namely, of seeming dogmatism.

Let me emphasize again: any well-stocked library has all the resources needed for any reader needing to learn 'the other sides' (plural in every case!) of this or that debate.

B. ONE MAJOR DECISION: EDDINGTON'S 'TABLE #2'

Rutherford's view of the non-atomic atom

Contemporary views of the physical world range from a defiant rejection of all rejections of naïve-realist views of the physical world, to the outright denial of anything physical whatsoever. The view chosen — with great confidence — as the best interpretation of all the available evidence is the view of Rutherford, the one which Eddington thought should give an 'abrupt jar' to all of us when we take a moment to

The Wonderful Myth Called Science

compare our everyday ideas of hands, books, and every thing else physical, with the discoveries made during the past century.

Rutherford's idea came after many generations of earlier thinkers tried their hand at the science or philosophy of nature. It began when the imaginations of some ancient Greeks were fired up by Thales's bold claim that, not only is water *water,* but everything else is water as well. One thinker after another found reasons for being dissatisfied with Thales's original proposal and argued for what they thought was an improvement.

The reason for mentioning those Greeks is because at some point the idea occurred to one of them that visible bodies are made up of invisibly tiny ones. Then it occurred to another person that there are different kinds of invisible ones. And so it went until it was proposed that the visible changes in visible bodies can all be accounted for if there are four types of tiny, invisible bodies. Since the tiny bodies were deemed indestructible — oranges can be cut and re-cut only a finite number of times — they were called "uncuttables" or, more commonly, atoms.

Aristotle was impressed by the theory, but when he reflected on his own legs, torso, shoulders, arms, neck, and head, he was even more impressed by the fact that he was one, single, whole being, not zillions of invisible little atoms somehow sticking together like tiny magnets or burrs caught in Velcro. It's the same with us. We don't refer to ourselves as "we," but as "I." (Unless we think we are royalty or newspaper editors.) Aristotle did what some modern thinkers do. He said that the visible bodies are the real, unified things, but that in some way the four elements or chemicals were 'virtually' present. Not actually. Only virtually.

Descartes' great discovery about human knowledge delivered the *coup de grâce* to Aristotle's worldview by showing that no bodies whatsoever are visible or tangible, not even our own. He was the first to notice and use the implications of what we today refer to as phantom-limb sensations. If Copernicus threw Aristotle's theories of motion into chaos, Descartes' physics and physiology, particularly his nascent neuroscience, cut the ground from under Aristotle's 'abstraction' theory of knowledge, thus completing the overthrow of Aristotle and, it should be added, the foundations of Saint Thomas Aquinas's two-sciences worldview.

However, Descartes' positive achievement was even greater. Having shown the inadequacy of Aristotle's natural philosophy and psychology, he presented in *Meditations on First Philosophy* a brilliantly conceived, unifying worldview that made use of the best ideas from medieval and modern thought. His *Meditations* are focused like the rays of the sun on attempting to answer one central question: "How can we know anything at all for certain?"

Descartes' own natural philosophy or philosophy of nature was inadequate. But he had provided a clear framework for the increasingly fast-paced discoveries pouring in from all over Europe. Those interested in the physical world took it for granted that, with reason to guide us, we can concentrate on trying to create more and more accurate picture-theories of what is going on out there, unseen, among those invisible and intangible bodies.

Once more, the library has abundant resources for anyone interested in the history of discoveries which led to Rutherford's 1911 astounding breakthrough, the one that suggested that uncuttables or atoms are not uncuttable or atomic. Also, once more, popularizations by experts who know the history, the experiments, and the various proposed models, are ideal resources for new learners still blithely unaware that their everyday thinking is riddled with insuperable inconsistencies. Here, culled from several instructive popularizations are statements supporting the decision to believe Rutherford was right. Start with him.

> ...when I hear today protests against the Bolshevism of modern science, I am inclined to think that Rutherford, not Einstein, is the real villain of the piece. When we compare the universe as we had ordinarily preconceived it, the most arresting change is not the rearrangement of space and time by Einstein, but the dissolution of all that we regard as most solid

into tiny specks floating in the void. That gives an abrupt jar to those who think that things are more or less what they seem. The revelation by modern physics of the void within the atom is more disturbing than the revelation by astronomy of the immense void of interstellar space.

The atom is as porous as the solar system. (A. Eddington, *The Nature of the Physical World*, p.1)

The space in the atom outside the nucleus is enormous compared with the size of the nucleus, or with the much smaller size of an electron. In the atom of hydrogen the single electron is near the outer rim of the atom. If its nucleus were enlarged to the size of a baseball, its electron would be a speck about eight city blocks away. Actually, of course, this atomic distance is small. The diameter of a hydrogen atom is nearly 1/200,000,000 of an inch; in other words, 200,000,000 hydrogen atoms could be placed one next the other in an inch. Relative to the nucleus or to the electron, however, the atomic space is prodigious. (Selig Hecht, *Explaining the Atom*, p.64)

From that day in 1911, when Rutherford described the inside of the atom, our whole idea of matter has been changed. The atom, formerly likened to a solid billiard ball, has become a transparent sphere of emptiness, thinly populated with electrons. The substance of the atom has shrunk to a core of unbelievable smallness; enlarged a thousand million times, an atom would be about the size of a football, but its nucleus would still be hardly visible—a mere speck of dust at the center. (Otto R.Frisch, *Atomic Physics Today*, p.13)

The new atomic model was definitely planetary. The surprising thing of the planetary model was how small the nucleus appeared. If the golf ball-sized atom was once again inflated, this time to the size of a modern sports arena or football stadium, the nucleus of the atom would be the size of a grain of rice. (Fred A. Wolf, *Taking the Quantum Leap*, p.75)

Quintalism & Subatoms

Now how do we know this is true if we can't see it? What proof have we that matter is made up of these quintillions of infinitesimal particles? Robert Millikan, one of the world's most noted physicists, said, "We can count the exact number of molecules in any given volume with more certainty than we can count the population of a city or a state." An atom is the smallest part of an element that can exist either alone or in combination with other particles. There are more atoms of hydrogen in a pail of water than there are drops of water in all the oceans of the world combined. So small is the diameter of an atom that half a million atoms piled, one on top of another would not even equal the thickness of this page! The volume of the average atom is about 1.56×10^{-25} of a cubic inch, which means that there are approximately fifteen–thousand–six–hundred–billion–million–million atoms to a cubic inch. Of course, such a number is totally incomprehensible, yet in spite of its inconceivable minuteness, the atom is mostly empty space! Its entire mass is packed into its nucleus which, believe it or not, is one trillionth the size of the atom itself. This is very fortunate. If the atom were all nucleus without any space in it, a glass of water would weigh as much as a two-ton truck and you would weigh as much as half a dozen locomotives.... Because they are so incredibly close to the nucleus these electrons make approximately 10,000,000,000,000,000 revolutions around it every second... (Jerome S. Meyer, *The ABC of Physics,* pp.22, 34-35)

A bar of gold, though it looks solid, is composed almost entirely of empty space: The nucleus of each of its atoms is so small that if one atom were enlarged a million billion times, until its outer electron shell was as big as greater Los Angeles, its nucleus would still be only about the size of a compact car parked downtown.... Nor, to return to the old classical metaphor, does a cue ball strike a billiard ball. Rather... on the subatomic scale, the billiard balls are as spacious as galaxies, and were it not for their like electrical charges they could, like galaxies, pass right through each other unscathed. (Timothy Ferris, *Coming of Age in the Milky Way*, pp.288-89)

In every single drop of sea water, there are fifty billion atoms of gold. One would have to distill two thousand tons of such water to get one single gram of gold. If we magnify the atom to the size of a football, the nucleus would be but a speck in its center and the electron, still invisible, would be revolving around its surface. Similarly, if we picture the atom as large as New York's Empire State Building, the electron, the size of a marble, would be spinning around the building seven million times every millionth of a second. There is relatively more empty space in the atom than between the planets in the solar system. (Bernard Jaffe, *Crucibles: The Story of Chemistry*, p.83)

We can end with the May 1995 issue of *National Geographic* cited in Chapter 3. In "Worlds Within the Atom," J. Boslough tells the same story told in 1911:

For each proton in the nucleus, there was a negatively charged electron, gyrating around the atom's core at a distance 50, 000 times the diameter of the nucleus. If a hydrogen atom's nucleus were the size of a tennis ball, its electron would be two miles away. (J. Boslough, "Worlds Within the Atom," p. 654)

What do these passages prove? — by themselves nothing. Forty thousand Frenchmen can be as wrong as one New Zealander can be right. They all represent individuals' guesses about a physical world that none of them ever laid eyes on, and all of those guesses may be mistaken. I think they are not mistaken, but I cannot apodictically prove that. I take the existence of physical things on faith.

"Faith versus reason" must go! A stipulative ' definition' of faith.

Before going any farther, it is necessary to stipulate a far more useful definition of "faith" than is customary. Descartes began Meditation One with a picture which is invaluable for all of us who decide the

105

unexamined life is not for us. He reflected that his mind was absolutely stuffed with information, with ideas or thoughts about an endless list of things. How, he asked himself, would he ever be able to rummage through that vast storeroom he called memory and find one idea that he could not possibly be mistaken about? The analogy of looking for a tiny needle in a haystack isn't strong enough. But he succeeded in finding just such a thought. *"Cogito, ergo sum,"* or "I am thinking, so I must be existing." Neither of those giants of antiquity, Plato and Aristotle, ever found anything as certain.

The fact that we can understand Descartes' *Meditations* reinforces one of the major theses of this unifying theory. Everyday thinking is chock full of often tacit convictions, thoughts we take for granted without ever putting them into so many words. Children who understand "Do you hear me!", "Are you fibbing?", "You're wrong, there's no Santa Claus," etc., already have the raw material that the great psychologists, beginning with Socrates, have toiled to put into so many words. (Or would have, if words existed.) The raw material for understanding Descartes is our everyday thinking that uses such terms as believe, reject, doubt, suspect, take for granted, assume, ask, argue, reason it out, lean toward, suspend judgment on, feel certain, and . . .

And faith! The expression "faith versus reason" is one we have all heard far more than twenty times. The two go together the way that mom and apple pie, objective and subjective, large and small, true and false go together, namely, as sounds that make us think of opposites rather than of things that go together. But reason and faith should always go together. The best faith is assent justified by sound reasoning.

In order to be clear about what "faith" will mean in these pages, it is best to stipulate a precise definition that must be understood when precision is called for. Here, it will mean the self-conscious attitude of willfully assenting to an understood thought when evidence for the thought's truth is less than conclusive. Less than conclusive evidence can range from no evidence to nearly conclusive evidence.

The Wonderful Myth Called Science

So important is that idea that it needs to be repeated — we take something on faith when, even though my evidence for it is less than 100% conclusive, we still assent to it.

I am absolutely certain that I understand that thought. I am absolutely certain that my visual field this moment has, near its center, dark figures I refer to as words. I am not certain that anything material exists. I still assent to the thought that bodies exist, mostly because I grew up 'programmed' to believe in physical things and because I have never found any reason to doubt it, even if none of those physical things is larger than a subatomic particle. I take the existence of physical things on faith, therefore. But I am absolutely certain I understand that thought and assent to it. That is, I am not taking it on faith that I am here and now understanding this/these ongoing, part-less thought(s).

I also take the existence of other people on faith. My scale of certainties is, in many respects, parallel to the scale of certainties Descartes expressed in *Meditations*. I believe it is possible, thinkable, that I am deceived, both about invisible and intangible material things and about equally invisible and intangible other persons. But I have never once doubted that bodies and other persons exist, and I have no evidence whatsoever that they do not. Further, though it is utterly immaterial to me whether or not material things exist, the existence of other humans makes all the difference in the world to me. The difference is that I also believe that I have abundant evidence, positive evidence, to support my faith in 'other minds.'

Naïve realism vis-à-vis common sense

Finally, a word about naïve realism. It will often be used interchangeably with common sense. That is because other writers use it that way. But, in order to emphasize that five-year-olds also have a deep-rooted but tacit sense of truth vs. error, as well as a tacit sense of right vs. wrong, I have found it useful to use "common sense" for a broader concept and to subdivide that common-sense worldview into three parts.

Here, then, common sense will often include (i) a tacit sense of the difference between thoughts that are true and those which are false, (ii) a sense of the difference between right and wrong that is initially 'embodied' in the do's and don'ts of one's family, tribe, culture, etc., and (iii) naïve realism. "Naïve realism" as the third component of a more-inclusive common sense means, most particularly, the two-sided conviction that material things exist and that they are in fact the things we see and touch. Since we initially believe we can see people's hands, legs, etc., even our own face 'in the mirror,' we initially believe that we and other persons are material beings.

The use of "naïve realism" in the narrower sense is the foundation for what Einstein and Bohr refer to when they use "classical mechanics," "classical theory," and/or "classical physics." Their theories of relativity and quantum reality, on the other hand, clash in radical ways with everyday naïve realism. The latter, as Kant realized, is structured implicitly by Euclid's geometry, Aristotle's logic, and Newton's physics. Like Aristotle before him, Kant was trying to make explicit what was already implicit in all human thinking.

What are physical bodies like? If they exist, that is.

What does the earlier string of quotations prove? Nothing by itself; but the quotes serve three purposes. They show that I was not the first in history to discover the idea of subatoms. I had to discover it for myself, but I did so, it seems, by reading the words of (often) older discoverers. The idea of subatoms was 'in circulation' long before I first learned of it and decided it fits into this unifying worldview better than any of its rivals. Secondly, the idea of subatoms has seemed to numerous experts to be intelligible, no matter how initially incredible it is. How incredible? Rutherford said he found the results of his experiments as incredible as firing a 15-inch shell at a piece of tissue paper and having it bounce back and hit him. Finally, the reliance of theory, that is, of thought, on images drawn from everyday thinking is beautifully illustrated by the way these writers strain to out-do each other with analogies drawn from common sense. No one builds their concepts of baseballs, city blocks, bars of gold, 15-inch shells, tissue paper, and so on, from their concepts of protons, electrons, and empty space. Just the reverse is true.

The Wonderful Myth Called Science

Question; how do we acquire the common-sense, everyday concepts of those mid-range bodies that are utterly indispensable prerequisites for creating concepts of macroscopic things such as galaxies, solar prominences, planetary rings, etc., which we say are 'visible' only by using telescopes, and for creating concepts of microscopic things such as microbes, double-helix strands of DNA, single molecules, atoms, protons, electrons, and photons, things which we say are 'visible' — some of them at least — only by means of microscopes? Since not one of them is really visible (what we say is false), how do we acquire our concepts of them, including our concepts of telescopes and microscopes? Einstein's answer is clear. We must create all of our concepts. We do not get them by observing whatever realities we believe the created concepts are 'of.'

Subatoms offer a special problem. Even in our everyday thinking, we know our eyes, allegedly made up of huge quantities of subatoms, are far too large to see an individual proton, electron, or photon. Not even with a microscope, also made up of huge quantities of subatoms, can we hope to ever see a single subatom. The latter are, we know, very different from eyes, microscopes, telescopes, etc., i.e., from human-eye-level or mid-range bodies. We think we get our ideas of the mid-range bodies from sense observation, because we originally think we see bodies directly.

Here is the special problem. In the case of macroscopic things, such as galaxies, Saturn's rings, etc., that is, in the case of things we say are visible only with a telescope, we initially assume that we could see them without a telescope, with our naked eyes, if we could hop aboard *Starship Enterprise* and get closer to them. Then we would get out our microscope to see what is really there. But if, as the physicists quoted above would say, eyes, the sun, and all other bodies are really not single things as our singular-form names for them suggest, then the names are group names for zillions of tiny bodies that no one can ever hope to observe, not even with scanning tunneling microscopy (STM). So, what is it that makes the quoted experts believe in protons, electrons, their quantities, their sizes, the distances between them, etc., with such confidence?

Quintalism & Subatoms

The role of mathematics is a subsidiary one!

Mathematics! Every book on modern physics assures us that, were it not for mathematics, 'the language of nature' (or is it the language of science?), researchers would never have discovered the incredible quantum story. (Far and away the best popularization of the history of the discovery of the subatomic world is Banesh Hoffman's *The Strange Story of the Quantum.*) But how do we acquire our knowledge of mathematics? We create:

> Thus, for example, the series of integers is obviously an invention of the human mind, a self-created tool which simplifies the ordering of certain sensory experiences. But there is no way in which this concept could be made to grow, as it were, directly out of sense experiences. It is deliberately that I choose here the concept of number, because it belongs to pre-scientific thinking and because, in spite of that fact, its constructive character is still easily recognizable. (A. Einstein, *Ideas and Opinions*, p.33)

This is part of his reader-response comment on Russell's epistemology (a fancy term for theories about knowledge). The next paragraphs in Einstein's essay do two things. (i) They praise Hume's meticulous dissection of the naïve illusion that we acquire concepts, e.g., of never-sensed causes, from sense experience. (ii) They also warn against Hume's skepticism vis-à-vis 'metaphysics'; Einstein was as hopeful about learning nature's secrets as Hume was skeptical.

The above statement of Einstein's is more far-reaching than most readers realize at first. We are told that it is thanks to mathematics that physicists have been able to discover so much about the unseen mysteries of unseen nature. Now we have to learn how we learn the mathematics needed to do physics.

Before we learn how we learn mathematics, recall Eddington's comment about Rutherford's atoms vs. Einstein's space-time. Einstein's space-time is used as long as we are earth-bound and can only use telescopes to study the heavens. True, men have reached the moon.

110

The Wonderful Myth Called Science

Wow, a quarter of a million miles away. The sun is four hundred times farther and it's only one of the billions of stars in our local galaxy. The next closest of those local billions is about four light years away — each light year equal to about six trillion miles. The idea that our Milky Way, with its billions of stars, is one of billions of galaxies, some of which initially look like a single star, the way a flock of birds might seem from a distance to be one bird, is astonishing in itself. The idea that, once we get close enough to examine one of the rocks on the moon, that rock turns out to be, not one thing, but more individual subatomic bodies than there are galaxies, is even more astonishing. And they zip about in space as empty as the space between the planets.

> I have settled down to the task of writing these lectures and have drawn up my chairs to my two tables. Two tables! Yes; there are duplicates of every object about me—two tables, two chairs, two pens. This is not a very profound beginning to a course which ought to reach transcendent levels of scientific philosophy. But we cannot touch bedrock immediately; we must scratch a bit at the surface of things first. And whenever I begin to scratch the first thing I strike is my two tables. One of them has been familiar to me from my earliest years. It is a commonplace object of that environment which I call the world. How shall I describe it? It has extension; it is comparatively permanent; it is colored; above all it is substantial. . . . Table No. 2 is my scientific table. It is a more recent acquaintance and I do not feel so familiar with it. It does not belong to the world previously mentioned—that world which spontaneously appears around me when I open my eyes, though how much of it is objective and how much subjective I do not here consider. It is part of a world which in more devious ways has forced itself on my attention. My scientific table is mostly emptiness. Sparsely scattered in that emptiness are numerous electric charges rushing about with great speed; but their combined bulk amounts to less than a billionth of the bulk of the table itself. (A. Eddington, *The Nature of the Physical World*, p.xi-xii)

Of course, neither table exists. The first is a fiction, pure and simple. What Eddington names with "Table No. 2" is a concept. It is a group-concept. At most it is shorthand for zillions of subatoms as such. So, both tables are non-existent. The same is true of books. What, then, do readers observe? How, then, do we learn mathematics? Whatever it is!

C. WE LEARN NUMBERS LATER THAN WE LEARN THINGS

We must begin with a thought-experiment.

Premise; we do not use mathematics to discover physical realities. We use our concepts of physical realities to discover mathematics.

We come into this world belief-less. We — at least our bodies — are bombarded immediately by heterogeneous stimuli — light, sound, odors, heat, etc., coming from the environment where they are jumbled together. First, they are filtered by our five sense organs. Eyes pick out light, ears filter out sound, skin responds to kinetic energy, resistance, and so on. The different or heterogeneous stimuli, after being filtered, trigger non-differentiated or homogeneous electro-chemical chain-reactions conveniently lumped together under "nerve impulses." Only these homogeneous nerve impulses, traveling via afferent neurons, reach the brain, at which point they correlate with utterly heterogeneous things, namely, sensory experiences (Einstein's vocabulary). "Sensory experiences" is shorthand for colors, sounds, odors, tastes, hot flashes, chills, etc.

Every twenty–first century introductory text for general psychology courses has detailed descriptions of these processes. The details, all regarded as 'scientifically verified,' must be kept in mind, clearly and distinctly, since so much of what follows relies on certain of them as irrefutable premises, i.e., as the (parts of) physics that disprove (parts of) naïve realism.

If what has just been described is accepted as stated, then we are ready to return to the question which Descartes, Kant, and Einstein were all keen on answering. How do we acquire that vast complex of everyday

112

beliefs that we rely on for our later 'higher' learning? The question is a retrospective one. The reason there are so many different theories about how we acquire our ideas is because we acquire them before we have enough of them to notice what is happening. Every theory about our acquisition of knowledge, from Plato's to Einstein's, was an attempt to figure out how we learn without knowing the rules for learning.

Putting all that has preceded together, we can accept one more thesis into our grand unifying or unified theory. We have to do all of our learning from inside. Whether it is from inside our skull as neuroscientists today are claiming, or from inside our mind as Descartes claimed, is a separate decision. For now, it is enough to say that we must imagine ourselves emerging into the world belief-less, being bombarded by heterogeneous sense experiences, and miraculously learning unself-consciously to 'make sense' of that sense experience. It will make it easier if we begin by breaking the learning process down into parts.

For that, it is instructive to imagine that we are back in our crib. In our belief-less state. Without any solid clue about what is really going on around us. So bereft of knowledge that we have not the slightest suspicion that we are clueless!

We first create an inner map or model of where we live.

The initial stages of our infant-to-toddler learning are especially conducive to the later thesis that our earliest learning consists in the formation of endlessly complex, criss-crossing habits of memory associations.

As newborns, we began immediately what would be a months-long process of constructing an inner map or model of our outer environment. The process can be grasped easily by anyone who has ever attended a college built on a large, sprawling campus, with large irregular buildings. Initially, it is necessary to ask directions, look for signs, consult onsite maps with helpful "You are here" arrows. In time, we become experts, able to give directions to strangers who ask how to reach any part of the campus or a building. As infants, the whole universe was unfamiliar. But with enough experience we became accustomed to what we would see when leaving one room and entering another, when going outside, and later when walking from our house to the candy store at the corner. In time, we should know where in the Milky Way our local star is located, where our sphere spins in relation to the rest of the solar system, even whether there is a difference between the top and bottom half of this sphere. (Which is which?)

The language-sounds that significant others made in our presence were just extra sounds that we had to 'make sense' of. There are only two hypotheses to consider here. Either we came into the world knowing every one of the estimated 3,000 to 10,000 languages, so that we could immediately acquire culture via the one 'native language' used in our social environment, in which case we simultaneously lost our ability to understand the other thousands of languages. Or else, in addition to the inner-model of our environment formed from sights, sounds, smells, etc., tied by association habits with other memory-clusters, we had to acquire additional associations between those clusters and the 'spoken' sounds later grouped under the label "my native language." The second guess is more parsimonious, fits smoothly into more parts of a greater, more comprehensive worldview, and is selected here.

That is an extremely compact sketch of a major feature of Kant's and Einstein's theory of knowledge. For them, the first role of our conceptualized theories is to provide a rationally-intelligible basis for making better and better predictions of our future sensations. The works of Descartes' immediate successors, namely, Locke, Berkeley, and Hume, were a major source for discovering the complexities of the association-facet of our early — and late! — learning.

We create our abstract 'mathematical' concepts

Our focus here is on mathematics. How do we acquire our ideas of numbers that no one has ever seen or touched? In the passage quoted above, Einstein says simply that we create our ideas of integers or numbers. To a degree, he is right. He is right to the extent that there are no numbers floating about or drifting overhead, already 'out there' just waiting for observant learners to observe them. But what prompts us to create concepts of numbers? We need some stimulus to prompt us to create the concept of Santa Claus and science. What about numbers?

Since space and time are not infinite, we will ignore the fact that the experts on mathematics vehemently disagree about every question ever raised about their subject. They do not even agree in their answers to the simple question, "What does 1 + 1 = 2 mean?" The only way to begin here is to begin *in medias res* — fancy Latin for 'in the middle of the mess.' We can do this, because we already have the idea of numbers. The question here is "How did we get it?"

Plato jumped right in. He was dazzled by the power of pure thinking shown in 'doing mathematics.' His solution was the hypothesis that, as early as birth, we already know the essentials of mathematics. We bring them with us from an earlier existence. In order to get over the difficulty that neither he nor anyone else he knew could remember thinking about mathematics while lying in their cribs, Plato proposed another hypothesis. All newborns suffer total amnesia on their entry to this life. As adults, we may recover memories of the time we were Julius Caesar leading an army over the Rubicon or of the day we were burnt at the stake as a witch. There are hypnotherapists willing to assist us with memory recovery, regression. But the vast majority of us who have read Plato's works reject his hypothesis as soon as we learn about it.

Aristotle, after an early apprenticeship to Plato, set out to show that we get all of our knowledge, mathematics included, from the raw material supplied by the senses. At first, the idea seems plausible, particularly if we begin with geometry. We begin with what we can see, for instance a three-dimensional pyramid. Each of its sides has three edges. How perfect for explaining how we get an idea of a triangle!

Here's how: first we cut the idea of one side away from the idea of the whole pyramid. "Cut away" can be translated as "abstract," a word whose Latin ancestry is "pull from." In other words, the concept of a pyramid has a lot of parts the way the pyramid has, and we can pull out that one part, the sub-concept of a pyramid's side, from the whole concept, and play with just that part, all by itself. This first mental move goes from a solid to a plane, and the plane has only two dimensions. To get the rough concept of a pyramid's sides ready for doing abstract geometry, we abstract the area in the middle of the plane, so that only the bare edges are left. By doing this second mental move, we have edges with no thickness and have gone from a two-dimensional plane to lines with only one dimension. Taking just one, we make it perfectly straight, the way we would pull a string tight with our two hands. This gives us the idea of a straight line that is the shortest we can get it between two hands or points. Do this to all three edges, till we have three lines. Where any two of them meet must be . . . ? Not a solid. Not a plane. Not a third line. It's just a point. That explains why Euclid defined points as things with no dimensions. If they had any, they wouldn't be points.

Unlike numbers and geometrical figures, pyramids (we think) really exist. Aristotle therefore concluded that mathematics requires an even higher degree of abstraction than physics. The physicists do abstract. For instance, physicists would be interested only in how pyramids in general come into existence, change, and perish over time. That is why "in general" is essential. Physics is not about one pyramid, but all pyramids, not about one time-bound body, but about all changeable things viewed in abstraction from change and time. The contemporary idea that science seeks general principles or universal laws is the same idea Aristotle had.

The mathematician, though, cares even less about reality. What she looks for is a science about the purely imaginary 0-dimensional points, 1-dimensional lines, and 2-dimensional planes; in this case, triangular planes. We know that if we talk about triangles, we will be talking about things with three 1-dimensional sides or lines, and so on and so on. Wallace's book is an excellent illustration of the way mathematicians can go on and on and on.

The Wonderful Myth Called Science

We don't have to do that here, because our goal is to ask "How do we get our idea of a number?" If you have about twenty books by the best experts, you can begin looking for an answer to that question. What you will find will be either "We don't know," or "We can't agree," or in my favorite book[*] "We don't need to know, we just go ahead and do our mathematics." For many people, the idea of a number gets more confusing the more books they read about it.

For instance, it turns out that Aristotle is useless when we switch from geometry to arithmetic and pure numbers. Some will tell you that 1 is not a number according to Aristotle or Euclid. It's a unit. Since Aristotle thinks of quantity when he thinks of mathematics, and thinks of quantity as dealing with plural parts (of one whole), it makes sense to say that the unit is no number, since it is the whole or the unit. Only when we get to parts, do we have plurality. 2, therefore, is the first number, not 1. And so on.

But where do we get our idea of 1, of a unit, of a whole? Some people hold up a finger, say "This is one thing," and imagine we get the idea of 1 anytime we see just one thing. But that clearly won't do. Anyone can see that looking at a thumb is seeing a thumb, not the number one! Looking at a thumb, index finger, biggest finger, small finger, and in-between finger is not looking at all five of the first numbers. (Is 0 a number?)

In fact, we can apply the RED experiment here. When I look at my index finger, which of the following tells best what I see: a finger, an index finger, my finger, flesh color, finger-shaped flesh color, part of a hand, part of a human body, a descendent of a toe on a foot from which

---

[*] That is O. G. Sutton, *Mathematics in Action.* NY Harper, 1960

117

hands evolved, flesh, skin, a three-jointed body-member, matter, a physical object, a visible thing, a thing, part of me, or . . .? No wonder there is no agreement, not even on whether mathematicians should bother debating the issue, hence the "No one knows, why bother?" attitude of some.

Those who, unlike Einstein, think all of our ideas come from sensation tend to turn a blind eye to the facts that Plato, Aristotle, Descartes, and Kant saw so clearly. What facts? Here is an example. Every schoolchild knows that $1 + 1 = 2$ always. Never 3, 6, or 9 always 2. So-called radical empiricists, e.g., David Hume, can reason as well as anyone. From their dogmatic rejection of all but the law of association, they feel driven to draw the logical conclusion that $1 + 1$ may not always $= 2$. They logically deduce that, if $1 + 1$ does not have to always $= 2$, then sometimes it may $= 3, 6$, perhaps even 9. Perhaps it never will $=$ anything but 2 in this universe which is the source of our experience. But what (empirical) evidence can we invoke to rule out the possibility that in some other universe $1 + 1$ might $= 5$? Or 50?

Mathematical propositions, Hume suggested, should be viewed as relations between ideas. This guess permitted Hume to join Plato and Descartes in viewing mathematical propositions as eternally true:

> Propositions of this kind are discoverable by the mere operation of thought, without dependence on what is anywhere existent in the universe. Though there never was a circle or triangle in nature, the truths demonstrated in Euclid would forever retain their certainty and evidence. (D. Hume, *An Inquiry Concerning Human Understanding*, Sec. IV, Part I)

Before we ask, "Can $1 + 1 = 2$ ever be false?" we must go back and ask again, "How do we get our ideas of 1, 2, plus, equals, and so on?" Chapter III has already made the argument that we do not get them by looking at ink marks. But if numbers are invisible, not-look-at-able, what kind of things are they? Like Locke, Hume implies that they are simply ideas. But what are ideas? How did we get the idea of an idea? And how do any ideas get into our minds in the first place? Once more, half of

The Wonderful Myth Called Science

Einstein's answer is Sherlock's Elimination: we don't get them from sense experience. His positive solution is "We create them." Is that clear, though?

An aside! "Creation" is a dodge. Hume would have enjoyed poking fun at the idea of our minds creating anything. He had already laughed at the idea of minds. Who ever caught sight of one of them?! As for creating ideas of numbers, when did we begin doing it, and how did we know enough to create so many different numbers that seem to have an eternal, uncreated order to them?! Everyone will admit that 6 comes after 5 and before 7. What makes that eternally or forever true, certain, and evident? The provisional solution here must be this. Einstein's idea that all of our concepts are created by us should be regarded as a provisional hypothesis, to be used until something better is revealed to us. So long as we retain Einstein's thesis that we can use concepts we create to discover realities we do not create, the better formula is not "we create," but "we create to discover."

Numbers? We start with unintelligible-to-us, repeatable strings of sounds.

Back to our days as cradle-bound infants. Fast forward to the days a few months later when we already have the beginnings of an inner map or model of the outer world. In addition to inner scenarios of moving from one place to another, we add 'items' to the model. Gradually, we are ready to associate names (e.g., "mom," "dad," etc.) with the items. In time, we will have memory-content rich enough that we can begin grouping our ideas of those items into sets, classes, categories, kinds, species, etc. (e.g., "people"). But we still have not seen any numbers! When do we begin creating ideas of numbers? When do we begin to discover that $1 + 1 = 2$? What is it that triggers us to create and discover any mathematical concepts?

It would make things much easier if we could climb back into our crib and re-view our personal learning history. Since we cannot do that, we must look for clues in youngsters we are acquainted with. The first thing that seems to make sense is that children begin by learning to count. As was the case with the children who could recite "igneous fusion" on cue, if it was the right cue, but who had not the foggiest notion of what it meant, children initially memorize pure sounds. For most, it takes time to repeat "one, two, three, six, four, nine" in the correct order. But only a minority of the world's infant population ever does that, because only a minority memorize 'English' rather than 'Italian,' 'French,' 'Chinese,' or other strings of sounds. Later on, and many of us can remember this, we memorized the multiplication table and could readily say "forty nine" when we saw 7 X 7 = ? or heard the spoken counterpart to the ciphers. What did it mean, though?

It is one thing to know 'how to count' correctly and another to have some vague idea of what a number is. A thirty–five year project of testing college students by writing 5 on the board and asking "What is that?" has produced enough evidence to have a strong faith that, as if they are dogs trained to bark on cue, they will utter the spoken counterpart to "the number five." Students initially find it hard to believe that the greatest mathematicians disagree, not only on what numbers are, but on whether they exist at all. But they are also initially incredulous when told that great physicists disagree, not only on what gravity is, but on whether it exists at all, just as they are when told that experts disagree on what "meaning" means. The trouble is that, like the author of these non-pages, they complete twelve years of formal education without anyone ever telling them that numbers do not exist, gravity does not exist, and language does not exist.

How can that be?! 'Everybody' knows numbers exist— and gravity. —and language. Oh, right! As for the letters of the alphabet, the beginning of the illusion that they exist is identical to the way the belief in numbers begins. Why else does it take awhile till "a, b, c, f, h, e," etc., can be sung in the correct order? Why else were the Romans too confused to know that I, V, X, L, etc., are letters, not numbers?

Slowly, we settle into tacit believings.

Such is the nature of our minds that, once our initial Twenty-Question preparation has begun in earnest, it becomes practically a law of nature that what is 'heard' as the name for some thing is accompanied by a tacit assumption that there is a thing that 'has' the name. One name, one thing! That is why we begin to believe there is a Santa Claus, why we later believe there are such things as science, religion, philosophy, and . . . and numbers.

Our capacity for critical thinking begins the day we change our knee-jerk reaction to statements in which a new, unfamiliar word is used. Early in our learning career, our reaction is the question, "What's a giraffe?" In time, we may balance that off with "What does 'giraffe' mean?" But it is not until we begin asking "Do giraffes or anything else exist?" that critical thinking begins. Now, we begin to see that Ockham's Razor must be traded in for an ax, so extensive is the process of simplifying reduction going to be.

If concepts and names for them existed, we could finish here by concluding that we have concepts of numbers, as well as names for the concepts of numbers. We could then leave it an open question as to whether or not numbers as such exist.

Not here, however. It is not an open question here. Numbers do not exist. Concepts do not exist. Language as a whole, including words and names, does not exist. None of those things are on the quintalist list of what exists. We do hear sounds. We do see figured black-cipher areas against a white ground. And we can understand complete thoughts about concepts, language, numbers, and so on.

And none of those thoughts exist in a vacuum. They cannot exist in isolation from someone's worldview or personal philosophy or belief-system. Einstein made that point in his comments on Russell's epistemology.

> In order that thinking might not degenerate into "metaphysics," or into empty talk, it is only necessary that enough propositions of the conceptual system be firmly enough connected with sensory experiences and that the conceptional system, in view of its task of ordering and surveying sense experience, should show as much unity and parsimony as

possible. Beyond that, however, the "system" is (as regards logic) a free play with symbols according to (logically) arbitrarily given rules of the game. All this applies as much (and in the same manner) to the thinking in daily life as to the more consciously and systematically constructed thinking in the sciences. (A. Einstein, *Ideas and Opinions*, p.34)

". . . it is only necessary that enough propositions of the conceptual system . . . and that the conceptional system . . . should show as much unity and parsimony as possible." No one could tell the difference between "*Guten morgen*" and "Go to hell" who did not have a system of beliefs as a context. Contexts are not out on the page or surrounding the spoken sounds. One's mindset is the context. Our divergent mindset-contexts shape our divergent responses to what we hear and see.

D. APART FROM REALITIES, MATHEMATICS IS PURE IF-THEN THINKING

Three all-inclusive, unifying questions.

To have a complete and completely true Grand Unified Theory, we need the answers to three questions. They are:

1. What exists?

2. What do existent things do?

3. Why do they do whatever they do?

The third question is moot if the answer to question two is "Nothing." The second question is moot if the answer to question one is "Nothing." That is why the answer to question one is the foundation for everything else.

Descartes' revolutionary reductionism contributed to a major simplification with respect to the first question. Once we are old enough to have a sufficient basis in concepts and vocabulary, we tend to expand our list of existent-things with utter naiveté. At first, we name every four-legged animal "doggie," but it is not long before it grows into distinct

122

ideas-plus-names for dogs, cats, horses, cows, rabbits, sheep, goats, etc. Reports from environmentalists tell of the numbers of already-lost species and the number of now-endangered species, but always against the backdrop of the thousands of species known to biological researchers. Our early Twenty-Question grouping of things-that-exist into minerals, vegetables, animals, and humans, is about as extreme a reduction as our everyday thinking permits.

Descartes' reduction of everything into extended bodies and part-less minds made it possible to take seriously the contention that every body is really nothing but atoms, or nothing but subatomic particles. By eliminating the need for distinct essences and forms, we can concentrate on discovering the laws 'governing' the only two things those tiny bodies can do: remain at rest or move. Max Born told us that "rest" is a word for something that does not exist. He overlooked the fact that "motion" is in the same category. Rest and motion are fictions, nouns created from verbs used in sentences that answer the second question, which is not "What exists?" but "What do existents do?" Bodies stand still or they move. Period!

The same type of analysis applies to minds. Minds think, reason, doubt, imagine, remember, anticipate, decide, and so on, but those doings are not things. When we convert the verbs to nouns, e.g., thought, deduction, creation, memory, etc., we are liable to add them as imposters to the list of existent-things. Fictions are handy. They are unavoidable. But we must master them, lest they master us. (A quick aside. Library resources, which begin with Parmenides who reduced things to one thing and Heraclitus who reduced doings to one doing and which end with 'Being' philosophies and 'Process' philosophies, will reveal how hard it is to avoid derailment while following out the preceding trains of thought.) Once we understand Descartes' brilliant simplification and decide it is the road to take in pursuit of wisdom, we must be careful to neither make the list of existents too short nor sneak eliminated things back onto the list.

In relation to the unifying worldview proposed here, Descartes' dualist or two-item list is three items short. Both materialist and idealist monists are four items short. Nothing less than persons, the thoughts they think,[*] the sense-data they experience, the copy-images of sense-data they can become more explicitly aware of, will suffice. Nevertheless, for reasons already noted, a quintalist such as this author will add subatoms to the list of what-exists.

A major clarification: How to 'reduce' concept-objects to thoughts.

Back to everyday thinking which might start like this: "I'm five years old. It's almost Christmas. I don't ask so many questions anymore, because I'm getting used to things. I live in a pretty big city with lots and lots of people who live in lots and lots of houses on lots and lots of streets. I've been in lots and lots of stores. They're all full of shelves with lots and lots of different things on them. When I go places, I don't get scared so much anymore, 'specially when I'm with my mom or dad. They're people, and I can tell people from dogs. I'm a little scared of dogs, so I stay away from them. I have a box full of old toys that I used to like, but not so much anymore. I know Christmas is coming. I can't wait for Santa Claus to bring me some new ones. Sometimes I can't tell what my mom and dad are talking about, but mostly I can. And I'm

----

* The word "thought" is as treacherous as "same." It can be used for an act, for what we do, as in "I thought you had gone." Or it can be used for the meaning of a complete sentence, that is, for the full object which we understand, as in "'that you had gone' is what I thought." Concepts do not exist as such, only the complete thoughts which we understand. Nothing but that principle will solve as far as possible the otherwise inescapable labyrinths or webs in which our theorizing becomes lost or entangled. (Can't you understand the complete thoughts, each true, false, or in between, that go through your mind as you read?)

getting pretty good at talking, like asking if I can have some more dessert, or asking if I can stay up a little longer instead of going to bed right away."

What is needed is the right model for analyzing such youthful thinking.

I understand thoughts. This is the only model that includes the essentials of all that and all complete-thought thinking. I am the agent. I do such things as see, hear, feel, and so on, and I also understand thoughts. It's the one me doing all those acts. When I go to sleep, I may stop thinking, at least when I'm not dreaming. But if I do stop, I don't stop existing. So I must think of the act of understanding as distinct from me. The act of thinking or understanding can be thought of as pretty much the same from moment to moment while I'm doing it. But the understood-thoughts are different. It's not one same thought that I understand over and over, as might be the case if I were reading "A rose is a rose is a rose is a rose is a rose is a rose is a rose" on thirty–five lines on every page of every book I ever read. I understand thoughts. A single I performs an ongoing act vis-à-vis different thoughts. I understand thoughts.

I do not understand Santa Claus. I do not understand my hands. I do not understand a's, b's, c's or p's, q's, and r's. I understand complete thoughts. They can be thoughts about Santa, hands, and everything else under the sun. If I see "red" or "1" all by itself in the center of an otherwise blank page, I do not understand red or the number one. I can think an almost endless number of thoughts about what I see. I never understand anything except thoughts which — here is the best way to put it — 'come to me.' I don't see or observe where they come from. And each thought is a single thing with no parts to it. The sentence correlating with the thought is broken up into words, at least for us. Theancienthebrewsrantheirwordstogetherlikethis. But the thought is a changing, part-less whole, not broken up.

It is convenient to pretend the successive thoughts are distinct, but the meaning of one carries over to the next in a way that can be described by a sentence that just goes on and on the way this one will keep going on and on so that you will get the idea that thoughts run on together so clearly that you will stop thinking that there is a clear and sharp break between sentences or rather between thoughts because thinking that way is a disastrously bad thought habit that you must get over and that you can get over if you make up your mind to keep working at it in the different ways being suggested in this text and I'll stop here because you get the idea. Right? (Right what? Right vs. left? A right to freedom of thought? The right way to think? Right vs. wrong? Lots of meanings of "right," right? How do we ever get so many ideas for one word such as "right"?)

It was probably inevitable that the first psychologist-thinkers who thought about what went on in their mind would believe in concepts. Our attention, when we are thinking, is on what we are thinking about, not on the thinking we are doing. Once we are old enough, we think the world is made up of individual things: individual trees, people, dogs, and so on. Images and concepts must be of individual things, therefore. What else could they be of?

Introspection seems to confirm that ancient tradition. As Hume noted, the easiest things to notice when we 'look inwards' are the images of individual trees, people, dogs, and other things we have seen. When we notice that we can use one word, "person," to refer to mom, dad, auntie, uncle, big sister, little brother, and so on, it is natural that we will conclude that the abstract concept named by "person" is distinct from different images of different persons. We construct a theory about thinking to fit our ideas about the things in the world we've seen and formed images of.

Only when our years of trying to use that old tradition to understand the relation of thought to things, especially of concepts to images-plus-sensations which come between thought and physical things, suddenly begin to give way to the realization that we understand meanings and that the meanings are thoughts, not a separate meaning for each separate word, but the much-at-once meaning of all the words . . . all the words in the sentence, contextualized by the meaning of all the sentences in the

The Wonderful Myth Called Science

paragraph that preceded it (the latest meaning), all the words in all the sentences in the chapter, the meaning of all the words it would take to express the vastly complex network of beliefs that form the silent backdrop for the focal thought correlated with this sentence, then the next, and so on . . . only then, when that thought 'comes to us' and we then work out the details of the rest of our system, only then are we in a position to see how to reduce concepts to complete thoughts.

Only then, finally, are we in a position to recognize how much false thinking we have been doing all our lives, but without ever realizing it. All of it was similar to the way we were only imagining that Santa Claus existed, since our thought that correlates with "Santa Claus exists" was false all along.[*]

Once our eyes have been opened we can then grasp what is meant or what ought to be meant by "pragmatic fiction," "theoretical construct," or "logical construction." We deliberately pretend. A hypothesis should begin as "Let's pretend that . . ." Let's pretend that my hand is made up of tiny atoms too small to see. We test our pretendings the way we test other guesses, e.g., "I'll bet she's home by this time." We test it against our sense-experience, both our memory of past experience ("You're right, it doesn't usually take her long to get home from here") and our future experience ("Let me call her to see if she got home alright"). We have Huxley's word (which we read) for it. Everyday 'science' is the only kind of 'science' there is; except a lot of everyday science is false. So are vast

---

* No, Francis Pharcellus Church, there is no Santa Claus, and it was a lie you told Virginia in your 9-21-1897 editorial in the *New York Sun*. And it is a still a lie to say "Yes there is a Santa Claus" when you know the listener will think you mean the toy-shop operator at the North Pole, not Saint Nicholas, that is, when you intend the other to be deceived about what you are thinking.

127

tracts of later science. But much false science is useful, even indispensable.

Mathematics as a tool used to think with precision.

Recall the reason for inquiring into what 'mathematicians' do. It is because everyone assures us that it is thanks to mathematics that physicists have been able to discover so much about the unseen mysteries of nature. No mathematics, no science! (Why else do the 'soft sciences' make certain that we take note of their use of quantitative analysis, i.e., mathematics.?) But there is a huge problem here.

Here's the problem. For a century or so, non-mathematicians have been told that knowledge is for those who know the mathematics of relativity and quantum theories. For the rest of us faith must suffice. We lay folk must trust that the experts know what they are talking about, even though our good sense tells us that their 'revelations from on high' are nonsense. "Einstein and Bohr are far, far out at the frontiers of knowledge, and we who are back here simply do not know enough to judge them" expresses an attitude that is common among far out physicists.

But that attitude has things backwards. No one can know what they are talking about unless they know what there is to talk about. We need the ideas we group together under "mathematics" in order to be precise about how many things there are. For instance, do one proton and one orbiting electron add up to two things, one proton and one orbiting electron, or to one thing, an atom? If an electron makes complete trips around the proton, we want to know precisely how many trips. Since each trip takes a certain amount of time, we want to know precisely how much time, or — if we pretend continuous time can be sliced into segments called seconds or nanoseconds — we want to know precisely how many seconds it took for the electron to make the aforesaid many trips. And so on. Eddington's famous example of the elephant sliding down the hill was an amusing way of presenting the same principle about the relation of 'mathematics' to the doings of bodies. Once we get the number-answers, we perform simple arithmetic or complex calculus with the abstract numbers, then translate back into everyday thinking. Since light travels 186, 000 mps, and since the sun is 93, 000, 000 miles away, and there are 60 seconds in each minute, we divide 93 millon by 186 thousand and divide that answer by 60, which yields 8.33. Translated back into real talk, we say that it takes sunlight a little over 8 minutes to travel from sun to earth.

The first of 'the sciences' in time were the theorizings of the astrologers who began by making charts of the star-constellations. Over the course of time, they began to notice that most stars were stationary relative to each other (imagine stars as being embedded in a transparent shell that shifts about in predictable ways), but that a few stars wander in and out of the fixed constellations. By making charts with dots to indicate the different places where the wandering stars appeared each night, and then connecting the dots to make 'paths,' the star-gazers could divide the 'paths' into units and match those distance-units with the time-units (whether days, months, or years). Thus, in the same way that distances down here on earth were measured by units of time, e.g., by the number of moons (months) a journey took, the distances traveled by celestial bodies could also be matched with units of time. Some later astrologers, we may guess, decided to change their name to astronomers, in order to distance themselves from astrologers who dabbled in bogus theories about the relation of our birth to the positions of the heavenly bodies.

Precision? Aristotle was content to say that rocks go faster and faster as they fall. Galileo took a giant step when he asked himself, "Just how much faster?" So, Galileo pretended there are invisible lines, e.g., from the invisible point where the body started its fall to the invisible point where it stopped, then he pretended the invisible line was divided into invisible parts, then he tested, ruminated, re-measured, tested some more, until he finally discovered a 'mathematical formula' that would fit all falls of most bodies, but not of the ones in the sky. It was left to Newton to pretend the moon is just another old body that 'obeys' the same laws apples 'obey.'

Quintalism & Subatoms

Convert declarative sentences into hypotheticals

The "would fit" is crucial. Most bodies seem to be at rest. The grains of sand at the world's shores move, but spend a lot of time just sitting. Boulders wait at the edge of a ledge for years before they fall. Trees are rooted in place. Houses have fixed foundations, and so on. Galileo's law of acceleration should be seen as hypothetical. If a body falls, and if there is no friction to slow it down, it will be going 32 feet per second faster at the end of each second than it was going at the beginning. But only if it is close enough to the earth's surface and if we pretend that what's underneath the nearest part of the surface of the earth has as much mass as the average mass of what's under other parts of the earth's surface. If-then. If-then. All if-then's.

Applied? Unless protons exist, it makes no sense to say there is only one of them among the two subatoms we call "a hydrogen atom." Unless electrons exist, it is pure make-believe to say that an electron completes "approximately 10,000,000,000,000, 000 revolutions" every second. And so on. This paragraph contains the key to understanding why Einstein held out against Bohr's interpretation of quantum theory until the day he died. They disagreed about what-exists, period! They did not disagree just about the states of what's out there, but on what's out there, period! That is, about what there is that might have a state or states.

Bertrand Russell's crystal clear capture of mathematics vis-à-vis physics

Two 'arguments from authority' — or two quotes, one argument — will introduce further clarifications. The first is from a Nobel Prize winner, Richard Feynman, who enjoyed teasing 'pure mathematicians.' The second is from a co-author of *Principia Mathematica*, one of the classic works on mathematics. Both illustrate the thesis here, namely, that mathematical manipulation of ideas and symbols is a mind-game comparable to video-games. Till we know what our numbers 'apply to,' pure mathematics tells us nothing about what-exists.

> ... I said there are never any surprises — that the mathematicians only prove things that are obvious.

The Wonderful Myth Called Science

Topology was not at all obvious to the mathematicians. There were all kinds of weird possibilities that were "counterintuitive." Then I got an idea. I challenged them: "I bet there isn't a single theorem that you can tell me—what the assumptions are and what the theorem is in terms I can understand—where I can't tell you right away whether it is true or false.

It often went like this: They would explain to me, "You've got an orange. OK? Now you cut the orange into a finite number of pieces, put it back together, and it's as big as the sun. True or false?"

"No holes?"

"No holes."

"Impossible! There ain't no such a thing."

"Ha! We got him! Everybody gather around! It's So-and-so's theorem of immeasurable measure!"

Just when they think they've got me, I remind them, "But you said an orange! You can't cut the orange peel any thinner than the atoms."

"But we have the condition of continuity: We can keep on cutting!"

"No, you said an orange, so I assume that you meant a real orange."

So I always won. If I guessed it right, great. If I guessed it wrong, there was always something I could find in their simplification that they left out. (R. Feynman, *Surely You're Joking, Mr. Feynman!*, p.70.)

Pure mathematics consists entirely of assertions to the effect that, if such and such a proposition is true of anything, then such and such another proposition is true of that thing. It

131

is essential not to discuss whether the first proposition is really true, and not to mention what the anything is, of which it is supposed to be true. Both these points would belong to applied mathematics. . . If our hypothesis is about anything, and not about some one or more particular things, then our deductions constitute mathematics. Thus mathematics may be defined as the subject in which we never know what we are talking about, nor whether what we are saying is true. (B. Russell, "Mathematics and the Metaphysician," sec.1)

What do we mean when we say there are thirty apples in a bag? Well, why not this: "If we start reciting our memorized string of sounds and make sure to say only one sound each time we pull one apple out of the bag, we will run out of apples right before we would have said 'thirty one'"? That's a favorite of many recent mathematicians who love 'matching' individual items in one group with individual items in another one. That won't do, because it's not the whole story. "There are thirty apples in the bag" makes sense only within the mindset-context of an entire worldview. There are no isolated thoughts. We must get in the habit of noticing that, in the back of our minds, we are using "thirty apples" as part of a way of thinking that, out of the indefinitely large number of different things making up this universe, we are thinking about only some of them, apples. The addition of a precise number only tells us how many of the 'some' we are thinking about. Etc., etc., etc.

What do we mean when we say "Don't think about apples, but only of the number thirty"? The answer here is, "There is no number thirty!" Numbers do not exist. But, in the same way that we can pretend that Santa Claus exists, we can pretend that numbers do. So, the number thirty is the one after twenty–nine and before thirty–one. Or is it five sixes, one hundred fifty divided by five, or only thirty times one? Regardless, it can only be those things in English. In Rome, thirty was XXX, something really different. But if you think hard about thirty being five sixes, you might begin to wonder whether it might not really be six fives. That is, if you try to connect the number-names to sets, you have to

132

decide whether thirty means five sets-of-six or six sets-of-five. Etc., etc., all the way to infinity and beyond.

For us whose goal is to find out the true and complete answer to "What exists?" the important thing is to not be bamboozled by the insistence that the royal road to reality is a profound understanding of mathematics. The royal road to reality is a profound understanding of what-exists. And people use very different numbers for their answer. Twenty-question people say "At least 4 genera, but an indefinite number of individuals." (N. B. Indefinite and infinite are not synonyms; the idea of infinity is the idea of a number that can't exist; we can't even get started counting toward a real, infinitely large number, because there isn't even one number to get started with.) Dualists say 2 essential types. Monists say 1. Quintalists say 5.

The point must be made crystal clear. Mathematicians disagree on their answers to "What are you talking about?!" for the same reason Aristotle and Descartes disagreed on what exists. The basic question behind all questions is "What exists?" Russell, who co-authored the massive three-volume classic, *Principia Mathematica*, pointed in the right direction when he wrote, "We 'pure mathematicians' don't know what we are talking about, because it could be anything or nothing; so it follows that we can't say whether what we are saying is true or not." Those who simply enjoy abstract thinking and have no desire to narrow themselves to thinking only about apples or only about photons or only about the rate of acceleration of falling bodies, are in fact talking about things that do not exist.

Or else, they are saying something that makes perfectly good sense, even though it is utterly mysterious why it does so. We can convert these declaratives into hypothetical if-then's. They are saying "If you have thirty of anything, then you will have one more than if you only had twenty–nine and you will have one less than if you had thirty–one." The ancients should not really have wondered what "one" means. They should have been decisive, said it is perfectly obvious (if anyone insists, admit that it is "intuitively clear") what having only one apple means, and then gone on to the really hard question, "Do apples exist?" and "If they do exist, what do we mean when we say an apple is really one thing?" Then, since people are more important than apples, the question is "Since I exist, am I right in believing I am one thing, period?!"

There is only one way to begin thinking seriously about mathematics. It is done the same way we beginners must begin thinking seriously about nuclear physics. It is done by reading dozens of books by experts who have read even more dozens of books by experts on mathematics. A. N. Whitehead's small *Introduction to Mathematics* will always be on my list, because it was my first introduction and because many of his observations about sense perception were instrumental in making me notice things that I had previously overlooked. It was not my first introduction to mathematics, obviously. I had the normal courses from arithmetic to logarithms, etc. I enjoyed the success of 'figuring' out mathematical problems. (Though certainly not as much as Andrew Wiles derived from solving Fermat's Last Theorem.) But that had nothing to do with my 'specialty,' philosophy. Since Aristotle's physics was perennial truth, distinct from the always-changing empirical sciences, and did not require mathematics, I had never had any desire or need to get 'introduced' to what I'd been doing in all those mathematics courses I'd taken!

E. FURTHER DECISIONS, BACKED BY ARGUMENTS FROM AUTHORITY

The real-life value of 'arguments from authority.'

Russell and Whitehead are said to have proposed in *Principia Mathematica* that logic and mathematics are, at their root, the same thing. Both are obviously severed from reality. If a + b = c is algebra, and if algebra is part of mathematics, it is clear that that formula has no obvious relation to anything real, not even numbers! Of course, we can pretend. As for logic, specialists in that abstract branch assure us that the argument from authority is a fallacy. They point out that, just because an Einstein uses Mother Hubbard toothpaste, that does not mean that it's as good as Crest, Colgate, or Tom's of Maine. (Of course, it also doesn't mean it's not equally good.)

The Wonderful Myth Called Science

But advertisers, who are more concerned with people and products than with whatever it is that pure mathematics and abstract logic deal with, rely on everyday psychology. They know that few things will increase sales of a product more than to link it (law of association) to an idea which, in consumers' minds, has a positive 'aura' to it or is associated with something else which has it. Scientists or wisdom-seekers do the same, whatever theories they may have about mathematics and logic.

For instance, George Berkeley (1685-1753) commented that some philosophers felt that an opinion was proven if they could cite a relevant passage from Aristotle. There are many people today who need no more than a passage from their sacred scriptures to convince them that something is true. For others, an opinion prefaced with "It is a scientific fact" or "It has been scientifically shown" instantly opens what would otherwise be a semi-closed mind. A quick visit to the library, especially to the section housing professional journals and reviews, will show how much reading all of the writers do. At times, the small print footnotes and bibliography, i.e., mostly arguments from authority, take up as much space as the 'corpus' of the article itself.

It will illustrate the need to be a student of more than 'one book' if another sampling of passages is attached here. Each of them relates to the complex relation between our views about what exists and the need to be not-gullible when reading what some authorities would have us believe. First, though . . .

Why more critical unified philosophies are better than less critical ones.

What we don't know can hurt us! Normally, what we don't know cannot hurt us. Not now, that is. If there is an asteroid headed toward the exact spot our planet is headed for, and if both bodies will arrive simultaneously at that spot exactly ten years from now, then what we don't know won't hurt us till then. That is plain common sense. "Let them enjoy the time left to them" is the reason doctors formerly used to justify not telling patients they had a terminal illness.

Quintalism & Subatoms

Today, when 'everyone' (somewhat of an exaggeration) touts the superiority of democracy, 'everyone' should know that democracy works only if the individual voting citizens are well-informed. Inasmuch as 'the argument from authority' is most powerful of all for ill-informed citizens, what the ill-informed or ignorant do not know or care to know or try to know can hurt them seriously in the future. Not everyone, of course. We who will die between now and the collision with the asteroid won't be harmed by it. If we vote for the wrong persons, but never have to go into battle, we won't be harmed when those persons get 'us' into a war. The question is, "What is it most important for us to know? ASAP?"

If 'all we know of mathematics' is a collection of vague ideas of lessons taken long ago and mostly forgotten, then we will have no defense against the defense of relativity absurdities already referred to earlier. A classic defense is that of A. d'Abro, in *The Evolution of Scientific Thought from Newton to Einstein*. After admitting that a deductive inconsistency would doom relativity theories, d'Abro wrote:

> However, it is scarcely probable that any such inconsistencies will ever be discovered by the layman, inasmuch as the theory of relativity is the work of mathematicians, for whom, more than for anyone else, accurate and consistent reasoning is the indispensable condition of research. (A. d'Abro, in *The Evolution of Scientific Thought from Newton to Einstein*. p. 212)

Only a familiarity with the endless disagreements among mathematicians, logicians, physicists, and every other group of experts, each an individual with a unique learning career that began with a common-sense philosophy similar to the common-sense starting point of every other expert in every other allegedly distinct 'meadow' (the variant for 'field'), will protect us laypersons from d'Abro's and many, many other writers' absurdities scattered here and there in any up-to-date library. That, of course, is a private opinion of this author. Therefore, when he can discover some expert who wrote something that makes eminently good sense to him he relishes it.

136

The Wonderful Myth Called Science

It is also his private opinion (there are no other kinds) that, unlike his earlier worldview, his present unified belief-system incorporates a just-right selection of the best private opinions of numerous experts who disagree vehemently with him on other issues —who do not all agree with each other on even one single issue except possibly when they are off-duty and relying on common sense! However, unless we have a well-enough-informed, unified theory of our own, we are at the mercy of absurdity-offering experts. Our best defense is familiarity with many experts.

(i) There is utter chaos among today's logicians

Immanuel Kant revered three disciplines: Aristotle's logic, the Greeks' mathematics, and Newton's philosophy of nature. He began the preface to the second edition of his famous *Critique of Pure Reason* with a claim that logic is a finalized science.

> That logic has already, from the earliest times, proceeded upon this sure path is evidenced by the fact that since Aristotle it has not required to retrace a single step, unless, indeed, we care to count as improvements the removal of certain needless subtleties or the clearer exposition of its recognized teaching, features which concern the elegance rather than the certainty of the science. It is remarkable also that to the present day this logic has not been able to advance a single step, and is thus to all appearance a closed and completed body of doctrine. If some of the moderns have thought to enlarge it by introducing psychological chapters on the different faculties of knowledge . . ., this could only arise from their ignorance of the peculiar nature of logical science. (I. Kant, *Critique of Pure Reason*, trans. N. K. Smith, B vii)

Kant's opinion appeared in print in 1787. The following century witnessed an explosion of new ideas about the way our mind works, or the way concepts relate to each other in judgments, or the way propositions relate to each other, and so on. In a partial reprint of a 1934 work, two writers summed up the changes that had taken place since Kant laid the foundations for the dogmas Quine so trenchantly criticized.

> Though formal logic has in recent times been the object of radical and spirited attacks from many and diverse quarters, it continues, and will probably long continue, to be one of the most frequently given courses in colleges and universities here and abroad. . . . But while the realm of logic seems perfectly safe against the attacks from without, there is a good deal of unhappy confusion within. Though the content of almost all logic books follows (even in many of the illustrations) the standard set by Aristotle's *Organon*— terms, propositions, syllogisms and allied forms of inference, scientific method, probability and fallacies — there is a bewildering Babel of tongues as to what logic is about. The different schools, the traditional, the linguistic, the psychological, the epistemological, and the mathematical, speak different languages, and each regards the other as not really dealing with logic at all. (M. R. Cohen and E. Nagel, *An Introduction to Logic,* 1962, p. vii)

Today's chaos, we will argue, lies in the contradictory worldviews that are used by those who have a module of theory labeled "logic" as part of their unified philosophy. Kant seemed to be making an inference: "Where there is consensus, there is science," the opposite of "Where there is disagreement about fundamentals, there is no science." Bad logic! Just because five people have five different views of something is not even close to a good reason for thinking that none of the five is right.

Add to that the obvious fact that, long before Aristotle wrote a set of works to explain his discovery of 'the' key to scientific reasoning, namely, immediate inference and the syllogism, humans had been reasoning as logically and illogically as Aristotle himself. Johnny who believes in Santa knows instantly that there's a problem when Susie says "There no Santa Claus." And he does that without having taken a course in logic. In fact, there are some of us who believe that, in many areas, Plato reasoned 'more scientifically' than Aristotle.

The reason is simple. In real life we use 'dialectic,' which should be described as 'modifying our stock of old opinions in a way that preserves as many of them as possible but still incorporates each new truth we learn about.' (That is the 'method' being used in this work.)

(ii) There is utter chaos among today's mathematicians

(This is not a superfluous addendum to the earlier discussion, because it relates to the important difference between Einstein and Kant.) In the preface to the second edition of his *Critique,* Kant also lavished praise on the other two 'finished' sciences, mathematics and physics. Like all who appreciate the fact that Newton's revolutionary *Mathematical Principles of Natural Philosophy* (1687) relied at every step of the way on precise measurements of masses, rest and motion, ratios between distances covered and times consumed, etc., Kant believed that the two sciences go hand in hand. The goal of his *Critique* was to reveal to the world why they go hand in hand and why they're apodictically certain. First, mathematics, "In the earliest times to which the history of human reason extends, mathematics, among that wonderful people, the Greeks, had already entered upon the sure path of science."

During the eighteen hundreds, however, Kant's view of mathematics, most especially Euclid's geometry, came under critical scrutiny. Many mathematicians concluded that Euclid's is not the only version of scientific geometry. Why Kant's view of Euclid is so significant is the fact that Einstein used a non-Euclidean, non-Kantian geometry for his general relativity. But the revolt went farther than just geometry:

> This book treats of the fundamental changes that man has been forced to make in his understanding of the nature and role of mathematics. We know today that mathematics does not possess the qualities that in the past earned for it universal respect and admiration. Mathematics was regarded as the acme of exact reasoning, a body of truths in itself, and the truth about the design of nature. How man came to the realization that these values are false and just what our present understanding is constitute the major themes. . . .

139

Quintalism & Subatoms

> Many mathematicians would perhaps prefer to limit the disclosure of the present status of mathematics to members of the family. To air these troubles in public may appear to be in bad taste, as bad as airing one's marital difficulties. But intellectually oriented people must be fully aware of the powers of the tools at their disposal. Recognition of the limitations, as well as the capabilities, of reason is far more beneficial than blind trust, which can lead to false ideologies and even to destruction. (M. Kline, *Mathematics, The Loss of Certainty*, 1980, Preface)

Recall that Einstein agreed with Kant's view that our mind creates the concepts we use to understand our streams of sense experiences. His two disagreements are these. First, he seems more confident than Kant that we can eventually figure out the true nature of nature in itself. That was directly linked to his confidence that our mind is not bound by a Euclidean straight-jacket. If non-Euclidean geometry fits our experience better, then it is a more reliable indicator of reality in itself. Together, the two disagreements led to Einstein's non-Kantian attitude towards Newton's physics.

(iii) Eddington's two tables and the utter chaos among today's physicists.

Here we reach the most important of the choices. Before me is a 1996 essay in a journal published by a prestigious 'scientific association.' The title of the essay is "When Science and Beliefs Collide." The essay divides the 'culture' into two groups, with scientists on one side and everyone else on the other. As if 'scientists' are not believers like everyone else! This version of the myth called science, one which is believed in by many people of the atheistic faith, is not one of the myth's wonderful versions. It shows an utter naiveté about the nature of thinking, everyday and specialist thinking, regardless of whose mind this particular version of the myth is lodged in. It displays the lamentable lack of self-knowledge reflected by the psalmist who praises God's mercy in a series of verses, one of which reads as follows: "Give thanks to the God of gods, for his mercy endures forever: Who smote the Egyptians in their first-born, for his mercy endures forever." The Egyptians were no better. They surely had prayers to match, just as those who reject science regard their own faith as superior wisdom.

Another interruption! Recall the 'faith vs. reason' or 'religion vs. science' myth. Where do we get those two or four concepts? Presumably no one would claim that faith, reason, religion, or science are visible or tangible. Einstein's is one answer. We create them. But what realities, if any, do the created concepts help us to discover? And how are we to verify or falsify any guesses or inferences about them? Is it the role of faith to prove that faith and reason really exist and are really distinct? Or is it the role of science to prove that faith really exists and is really distinct from reason? Or are these concepts ripe for being pruned with Ockham's Razor? They are not on the quintalist list of realities.

A crucial footnote fits here. Experts on Kant are aware that, in the preface to the second edition of his first *Critique*, he wrote "I have therefore found it necessary to deny knowledge, in order to make room for faith." (B xxx) This comment is made in reference to 'religious' faith in God, free will, and immortality. What is too often not pointed out is that Kant extends his denial of knowledge to external physical bodies as part of nature in itself. "However harmless idealism may be considered in respect of the essential aims of metaphysics (though, in fact, it is not thus harmless), it still remains a scandal to philosophy and to human reason in general that the existence of things outside us (from which we derived the whole material of knowledge, even for our inner sense) must be accepted merely on faith, and that if anyone thinks good to doubt their existence, we are unable to counter his doubts by any satisfactory proof." (A footnote to B xl) End of interruption!

More texts to support Kant's idea that everyone relies on faith (as defined above)

Here, to support the claim that there is not a single belief on which all 'scientists' agree, are some further arguments from authority. A major debate focuses on whether physicists' theories are false but useful theories or whether they tell us the truth about reality. This was the heart of the disagreement between Galileo and Bellarmine in the early 1600's. Both men agreed that future eclipses could be predicted either by Ptolemy's model in which everything revolves around the earth or by Copernicus' model in which the sun is at the center. Galileo held that Ptolemy's view, while useful for 'saving the appearances' (it fitted most of what was sensed), was false. Bellarmine insisted that the evidence was not yet conclusive.

In 1999, a conclusion in chemistry was reported. Researchers had finally 'observed' chemical orbitals or bonds in a molecule made of copper and oxygen. But had they? Here are two answers from two experts:

> "I want to emphasize," he [Dr. Spence] said in an interview, "that these pictures are not computer simulations. They are like photographs: true images of real objects." (M. Browne, "Glue of Molecular Existence is Finally Unveiled," in *New York Times*, 9-17-99)

> Let me now turn to the theoretical status and limitations of orbitals and why orbitals cannot possibly be observed. Atomic orbitals are mathematical constructs and strictly speaking are only genuine wave functions in one-electron systems such as the hydrogen atom. . . . The orbital approximation [elsewhere] is the basis of a great deal of work conducted in quantum chemistry, but here it is recognized that orbitals are mathematical constructs and do not possess any independent physical status. (E. Scerri, "Have Orbitals Really Been Observed?", in *J. of Chemical Education*, 11-11-00)

A lay reader may think that the only reason orbitals will never be observed is the obvious fact that, since what we see on the cover of the 12-31-99 issue of *Time* magazine is not Einstein but a photo, looking at a photo of an orbital is observing a photo, not an orbital. In fact, though, it is even more basic than that. The real bone of contention seems to be the bottom-line question, "Do orbitals exist?" (Do photos?)

The Wonderful Myth Called Science

More serious is the whole relation of atomic models to reality. In *Quantum Reality*, N. Herbert explains that he had found eight varieties of views on the issue. In his introduction, he mentions the most popular version, dogged faith:

> However, when I asked my teachers what quantum theory actually meant—that is, what was the reality behind the mathematics—they told me that it was pointless for a physicist to ask questions about reality. Best to stick with the math and the experimental facts, they cautioned, and stop worrying about what was going on behind the scenes. No one has expressed physicists' reluctance to deal with quantum reality better than Richard Feynman, a Nobel laureate now at *Cal Tech*, who said, I think it is safe to say that no one understands quantum mechanics. Do not keep saying to yourself, if you can possibly avoid it, 'but how can it be like that?' because you will go 'down the drain' into a blind alley from which nobody has yet escaped. Nobody knows how it can be like that. (N. Herbert, *Quantum Reality*, pp. xii-xiii)

That book was published in 1985. A year later, P. Davies and J. Brown interviewed eight leading quantum physicists to learn what they thought.

> A final thought and a note of caution; when we commissioned the interviews, several of our contributors (who shall remain nameless!) expressed the view that there is now no real doubt over how quantum theory should be interpreted. At the very least, we hope this book will show that there is little justification for such complacency. (P. Davies and J. Brown, ed., *The Ghost in the Atom*, p.x)

By keeping an eye open for reports of conferences and current research, it is possible to build up a body of reassuring confirmations of one's own intuitions. For instance, a G. U. T. hypothesis called "string theory" has stimulated enough interest among physicists to have become a whole new branch of physics. The trouble is finding evidence! For the

Quintalism & Subatoms

December 1987 issue of *Scientific American*, the contents-page editor introduced an article on string-theory as follows:

> Remnants of the infant universe, cosmic strings are massive, wiggling threads created in the second following the big bang. Their powerful gravitational effects may have caused the clustering of stars and galaxies that is evident in space—but before any firm cosmological conclusions can be drawn, theorists will have to prove that the strings exist.

In its 12-7-04 "Science Times," the *New York Times* reported on controversies regarding the theory. Comments ranged from one cosmologist's conclusion that string theory is "a colossal failure" to the opinion of others that the mathematics of the theory are "too beautiful to die." And so, while many young physicists are willing to devote their careers to string-theory research, critics bemoan the spectacle of so many brilliant minds wasting their careers "on a fruitless quest." Apropos of the lack of proof for the theory, one wag, several years ago, commented that "you can go on hot air only so long." Here is an example of how right Quine was:

> For years physicists have looked for the origins of string theory in some sort of deep and esoteric symmetry, but string theory has turned out to be weirder than that. Recently it has painted a picture of nature as a kind of hologram. In the holographic images often seen on bank cards, the illusion of three dimensions is created on a two-dimensional surface. . . . Just how and why a three-dimensional reality can spring from just two dimensions, or four dimensions can unfold from three, is as baffling to people like Dr. Witten as it probably is to someone reading about it in a newspaper. In effect, as Dr. Witten put it, an extra dimension of space can mysteriously appear out of "nothing." (D. Overbye, "String Theory, at 20, Explains It All (or Not)," *New York Times*, 12-7-04)

A major problem with string theory is that it requires more than three space dimensions and one time dimension, more even than space-time's

144

unified four-dimensions, possibly as many as eleven dimensions. The *Science Times* of 9-30-03 ran an article on Maria Spiropulu, a physicist who works in such laboratories as the Fermi National Accelerator lab. Her aim, and that of other researchers, is to find empirical evidence for a not-yet-proven fifth dimension. Only six or seven to go!

Our question here is: "Do space, time, or dimensions of any kind really exist?" Einstein's thesis is that those terms stand for concepts created by our imaginations, hence needing to be checked by sense experiences. Mathematics is useful, but only if we know what the measurements are meant to measure. Once imagination takes over and proceeds in the absence of reality checks, the results may gratify some physicists' aesthetic appetites, but they have nothing to do with knowledge of what exists. David Lindley, a Research Fellow in the Theoretical Astrophysics Group at the same Fermi National Accelerator Laboratory from 1983-86, wrote an early 1990's book entitled, *The End of Physics: the Myth of a Unified Theory.* A Science News advertisement led to my purchase of the book. It read as follows:

> David Lindley, a theoretical physicist and senior editor at Science contends that modern physicists are straying from true science in search of answers they can never hope to validate, such as the so-called Theory of Everything. He argues that physicists are blinded by the truisms and persuasiveness of mathematics. Lindley provides a solid overview of the developments in particle physics during the last century and concludes that we have not progressed from the time 100 years ago when physics was declared a dying field. At this point, he contends, physics is evolving into a mythological field.

In 1996, John Horgan, a former English major enamored of literary criticism theory, published a follow-up volume, entitled *The End of Science.* Horgan's book is a thoroughly human-interest work, not based only on theory, but on extensive interviews with dozens of the best-known thinkers of our time. The first person he mentioned was Sir Roger Penrose whose 1989 work, *The Emperor's New Mind*, had explored the relation between physics and consciousness. According to Horgan,

Quintalism & Subatoms

Penrose seemed to believe a Grand Unifying Theory was not impossible. The cover story for this month's *Discover* magazine is about Sir Roger's latest book, a one-thousand page work proving that Horgan's surmise was correct. Penrose's new book is *The Road to Reality*. The title itself expresses Penrose's confidence that unification is possible. *Discover's* introduction to the cover story begins thus:

> Is physics stuck in a rut? The question might surprise longtime *Discover* readers, who regularly hear about breakneck advances in fields ranging from engineered materials to atomic teleportation. But spend some time with the esteemed British physicist and mathematician Sir Roger Penrose and you may begin to wonder. ("Letter from Discover," in *Discover*, June 2005, p.27)

So, have we reached the end of physics? Have we reached the end of science as a whole? Or should we continue to hope that, despite the increasing disagreements, a Grand Unifying Theory is possible?

Our belief-system must not be too small, but have room for everything.

We must remind ourselves that "science" doesn't exist! Nor does physics in general or pyramids in general or numbers in any size, shape, or form. It is essential to keep our previous commitments within reach while thinking of other things. All of the preceding references to astrology, astronomy, physics, etc., must be seen as shorthand references to individual persons and their personal opinions.

Each and every one of those individuals was born belief-less like the rest of us, acquiesced into a starting-kit of common-sense beliefs, then read and thought their way to the later worldview which served as their mindset-context for specific opinions on specific issues. And those individual mindset-contexts varied radically.

But many of the individuals were models of truth-seeking. Disagree they might, but they each had a goal similar to that of the others' goals. The most famous case of G. U. T. seekers who ended with radically opposed mindsets was that of Einstein and Bohr. Their disagreement goes to the very core of the century-old logjam preventing the unification which Einstein spent the last years of his life working toward. (This year is the hundredth anniversary of Einstein's 1905 papers which were instrumental in developing both relativity and quantum theories.)

It had been years since I read Nick Herbert's book. When I pulled it down to copy the above passage, two things struck me. The lack of exact references left me disappointed (when one author cites a 'perfect' passage from someone else, we like a note in case we want to read it for ourselves). But he more than made up for that by his vivid report of the famous, thirty-year debate between the two giants of twentieth century physics. And a real debate it was. They met, they respected each other, they both knew all the formulas, all the experiments, all the angles. And each tried his utmost to change the other's belief-habits. In vain!

> Einstein's strategy was to confront Bohr with a series of thought-experiments which aimed to show that quantum theory had left something out. He did not attempt to show that the theory was wrong, but by demonstrating that it was incomplete Einstein hoped to open the door for what he called "elements of reality."

> As the winners tell the story, Bohr closed each of Einstein's loopholes, but in the minds of each the debate was never settled. Long after their arguments had ended, on the day Bohr died, his blackboard contained a drawing of one of Einstein's thought experiments. Bohr struggled with Einstein to the end.

> Einstein too never gave up. In his autobiography he expresses his final thoughts on the quantum reality question: "I still believe in the possibility of a model of reality—that is, of a theory which represents things themselves and not merely the probability of their occurrence." (N. Herbert, *Quantum Reality*, p.24)

147

Quintalism & Subatoms

In the next and final chapter, we will discuss the pivotal issue on which one might say the future of people's individual sciences or philosophies will depend. That pivotal issue is captured by the question repeatedly asked in the pages before this, namely, "What do readers observe?"

Whatever the answer is, it should now be more than apparent that the items in my overall view have been carefully selected from the wildly divergent theories which years of listening and reading have brought to my attention. The preceding samples have one conclusion. Familiarity with the entire range of wildly divergent views will, if nothing else, show us the need to begin making our own decisions. Others have played the game before us. Now it is our turn. No one else can play for us.

But we need not be afraid. Whatever we choose to believe, we can trust that it (a similar belief) was already adopted as the truth by some expert. We can examine the rest of that expert's belief-system — and their life and their critics — to see whether we do well to adopt it as our own. I have chosen many beliefs similar to those of Einstein. Like others, I think he was right about some things that Kant was wrong about.

Now, what do you think?

F. CRISS-CROSSING, CHOOSING-A-SIDE DECISIONS ARE ESSENTIAL.

Einstein was right to reject Bohr's conclusion about reality.

Sherlock's maxim tells us that the scientific method is two-sided. Eliminate the impossible theories. For those familiar with the Einstein-Bohr debate, it will be clear that the choice of Eddington's #2 table will be a rejection of Bohr's view about nature-as-it-exists-this-very-instant. (That Einstein himself fudged on 'this very instant' is a separate problem, dealt with later.) Let me explain.

It is overwhelmingly clear that Einstein's life-long, stubborn refusal to join Bohr and his followers was the direct, rigorously logical conclusion from his postulate, if we may call it that, of a reality that exists 100%

independently of us. The following passage is particularly clear. It was his direct reply in 1949, six years before his death, to "the detailed arguments of my highly esteemed colleagues Born, Pauli, Heitler, Bohr, and Margenau." All of those he named believed that a 'statistical quantum theory' operates with a complete description of undetermined reality. Einstein emphatically disagreed.

> What does not satisfy me in that theory, from the standpoint of principle, is its attitude towards that which appears to me to be the programmatic aim of all physics: the complete description of any (individual) real situation (as it supposedly exists irrespective of any act of observation or substantiation). (A. Einstein, "Reply to Criticisms," p.667)

". . . any (individual) real situation (as it supposedly exists irrespective of any act of observation)" is a 1949 repetition of his 1931 statement of principle in an essay entitled "Maxwell's Influence on the Evolution of the Idea of Physical Reality."

> The belief in an external world independent of the perceiving subject is the basis of all natural science. Since, however, sense perception only gives information of the external world or of "physical reality" indirectly, we can only grasp the latter by speculative means. It follows from this that our notions of physical reality can never be final. We must always be ready to change these notions—that is to say, the axiomatic basis of physics—in order to do justice to perceived facts in the most perfect way logically. (A. Einstein, *Ideas and Opinions,* pp. 259-60)

Bohr's 'Copenhagen,' statistical interpretation of quantum theory, when it is applied to anything irreducibly individual, whether it be the moon, an electron, or a photon, logically entails a rejection of Einstein's fundamental postulate or reality principle. And yet, Bohr insisted that Einstein was the one refusing to accept reality.

In his 1949 "Replies to his Critics," Einstein took note of the Bohr camp's standard objection to his faith in a fully independent-of-humans external world. As we will see, his defense of general relativity relied precisely on what he here rejects as 'a positivist' attitude. The passage cited above continues:

> ...irrespective of any act of observation or substantiation). Whenever the positivistically inclined modern physicist hears such a formulation his reaction is that of a pitying smile. He says to himself: "there we have the naked formulation of a metaphysical prejudice, empty of content, a prejudice, moreover, the conquest of which constitutes the major epistemological achievement of physicists within the last quarter-century. Has any man ever perceived a 'real physical situation'? How is it possible that a reasonable person could today still believe that he can refute our essential knowledge and understanding by drawing up such a bloodless ghost?" Patience! The above laconic characterization was not meant to convince anyone; it was merely to indicate the point of view around which the following elementary considerations freely group themselves. (A. Einstein, "Replies to his Critics," p.667. P. Schilpp ed. *Albert Einstein: Philosopher-Scientist*))

Einstein's reality principle or postulate is perhaps the most fundamental postulate of our everyday thinking. The unification proposed in the present book, naïve-realistically speaking, is emphatically in agreement with Einstein's principle. None of us thinks that the book we are reading exists only until we put it down and go off for lunch, or that it re-exists only when we return to resume our reading.

Einstein was right about the fatal flaw in the Bohr group's attitude. Believe it or not, they do not believe books exist unobserved. That conclusion follows logically from what Einstein very, very carefully refers to as "the statistical quantum theory." The word "statistical" is crucial. One of its opposites is "individual." That is the opposite which we invoke here. A second way to interpret "statistical" is to make it mean "undetermined," in which case its opposite is "determined."

The Wonderful Myth Called Science

The distinct concepts or meanings are usually sorted out by referring to cases in which we claim to be more sure of what a group will do than what an individual in the group will do. Insurance companies study deaths in order to learn how many individuals of a certain age and sex, e.g., seventy–two-year-old males, will die in the coming year. Suppose insurers predict that 5% of that group will be dead within the year. No one can pick out just which individuals will be among the dead group. In a certain mass of radioactive atoms, quantum theory makes it possible to predict how many of the individual atoms will break into parts or 'decay' in a certain amount of time, but no one can know beforehand just which ones will be among those that have 'died' or 'decayed.' The question is, "Why not?" Why can't we know?

But "Why not?" is a separate question. Unless there are individual atoms, it makes no more sense to ask "Why can't we be certain which atoms will decay in a given amount of time?" than it does to ask an unmarried bachelor "Why don't you quit beating your wife?"

Einstein insisted that it is simply a lack of information. For instance, if we had divine omniscience, we could pinpoint the individual seventy–two-year-old males who will die next year. (According to Einstein.) Similarly, if we knew every last detail about the individual atoms, we could pinpoint just which of them will decay at what precise future moment. Bohr's team disagreed, saying that it is not a matter of not knowing enough. There is nothing to know. At least there is nothing definite to know about individual protons or electrons unless and until we use light or electron beams to 'observe' their precise location or state of motion. As long as we do not try to observe them, the individual subatoms exist in a state of limbo or a shroud of mere probability. What happens is pure chance at the individual level, although — strange to say — we can predict what will happen to the group as such. But groups do not exist.

Further details about the thirty-year debate between Einstein and Bohr must be sought in the library. Bohr was right to say that we could learn every last detail about subatoms and still be unable to predict what will happen next to each distinct individual particle. Einstein was right to say that there is nothing imprecise about the individual subatoms. True, we disturb them by hitting them with light or electron beams, but if we do not, there is nothing fuzzy whatever about them, only about our knowledge. If subatoms exist at all.

Once more, that Bohr's team made reality depend upon us, the 'positivist' position Einstein rejected to the end, is not in any doubt. A small sample of evidence will show what Einstein rightly rejected as absurd. Bohr's view is a disguised version of Berkeley's argument that we cannot make sense of real bodies independent of us. The first 'argument from authority' is my personal favorite. It comes from a richly rewarding volume, *Philosophical Consequences of Quantum Theory: Reflections on Bell's Theorem*, edited by J. T. Cushing and E. McMullin. The other three samples will be restricted just to the titles.

> ...The questions with which Einstein attacked the quantum theory do have answers; but they are not the answers Einstein expected them to have. We now know that the moon is demonstrably not there when nobody looks. (N. D. Mermin, "Quantum Mysteries for Anyone," in Cushing and McMullin, p.50) "The Universe as Hologram: Does Objective Reality Exist or Is the Universe a Phantasm?", by M. Talbot, in the 9-22-87 *Greenwich Village Voice*. "Quantum Weirdness: Physicists are wondering whether a tree—or anything else—must be observed before it really exists," by M. Gardner, in the October, 1982 *Discover* magazine. "Quantum Theory and Reality: The doctrine that the world is made up of objects whose existence is independent of human consciousness turns out to be in conflict with quantum mechanics and with facts established by experiment," by B. d'Espagnat, in the November 1979 *Scientific American*.

Kant and Einstein both believed that there is a reality independent of us.

Einstein was clear. "The belief in an external world independent of the perceiving subject is the basis of all natural science." He also believed we are getting closer and closer to understanding that independent-of-us world. Kant re-defined "world" to mean the world as it appears or, in today's vocabulary, a virtual-reality world. But Kant also believed that there is, beyond the virtual-reality world, a reality wholly independent of us. We can have certainty about the laws governing the apparently-external, virtual-reality world's sensible appearances. But, even if we cannot have theoretical, deductive certainty about the really-external world, we still can aim for and achieve a reasonable faith about it.

Both Kant and Einstein will be interpreted here as agreeing on one further, certainly true conclusion: each of us must learn for ourselves everything we know about what is wholly external or outside our private stream of conscious experience. We must learn it from inside, because all of our evidence is inside. That is, all of our evidence for what is 'out there' must be found 'in here.' The foundation of everyday thinking is that there is a difference between our thinking in here and an independent reality out there.

What about space and time?

After numerous thinkers had done their best to understand space and time, Isaac Newton finally succeeded. He cut through the clutter and put into words the tacit, common-sense assumptions that frame our everyday thinking. (Every decent library should have handy anthologies to help readers learn about the wildly divergent theories that have been proposed about space and time!) Like Kant, we should welcome Newton's achievement. It must be part of our foundation for anything we hope to learn about subatom-sized bodies, if they exist.

We should also welcome Kant's great achievement vis-à-vis space and time. It does not seem to have occurred to Newton to doubt that something called space and something else called time were real, that is, existed independently of us. (He boasted about not relying on hypotheses.) It did occur to Kant. His great achievement was to 'bring space and time indoors,' that is, into our mind. We do not acquire the space-time framework for everyday thinking by directly observing a really external-to-our-experience world, spread out in real space and ever-changing in real time. Our minds are pre-programmed to create concepts of space and time to 'put order into' our spread-out, ever-changing streams of experience.

Kant, however, did not realize that concepts do not literally exist. There are, to be sure, 'faint copies' (Hume) of earlier sensed data. Those images are organized into a world-model by a vast network of criss-crossing associations. And, most importantly, there are complete thoughts. The model-imagery is used to decide which complete thoughts we understand are true, no matter what they are about.

Einstein erred as well. First, he tried to do what cannot be done, namely, to escape an absolute reliance on our original beliefs — complete thoughts — about the way bodies, if they exist, are spatially related. Space as such is neither Euclidean nor non-Euclidean. There's no space as such.

He also tried to do another thing that cannot be done, namely, to escape an absolute reliance on our common-sense beliefs about past, present, and future. Since all of space, if it existed, would be taken up by what exists right this instant, and now this instant, and now this one, and . . ., etc., there is no place in space for the past (which, we say, no longer exists) or for the future (which, we say, does not yet exist). In fact, there is not even any now, only the five things on the quintalist list.

Logical fictions and the "No concepts, only part-less thoughts" truth.

No matter how difficult it is to believe thoughts such as these, we can understand them. If we didn't understand them, we would not know whether or not they were difficult to believe. Depending on whether we prefer English or Latin, we can call that the Rorty Rule or the *Intelligo*.

We must distinguish between our thoughts about things and the things our thoughts are about. Eventually, we notice that some of the 'things' we believe in do not exist. Santa is a favorite here. In order to refer to the non-existent 'objects' or 'contents' of false thoughts, we coin the phrase "a fiction." Thus Santa and science-in-general are fictions. Or, we might prefer "mythical entity," or "construct," or "pragmatic fiction," etc. We can take over the model or picture the Greeks invented, and pretend that, besides things out in the real world, we have concepts or ideas of them in our minds. That way, we can divide our ideas into two or three imaginary piles. One pile has concepts we think match what exists out in the real world, the other has concepts we don't think have matching, real-world counterparts. A third pile would be made of concepts which we're not yet sure match or don't match.

Then we have to work to get over the myth that concepts as such exist. That sentence expresses a complete thought, not a string of concepts. Only by finally learning how to understand all of this in terms of understanding complete thoughts (the only things that are true or false) as opposed to creating concepts, then 'putting them together' into affirmative and negative judgments, can we hope to reach the pinnacle of insight.

Space and time are not infinite — so it is time to wind up this chapter lest it take up too much space.

G. SOME CONCLUSIONS. OR PREMISES FOR WHAT FOLLOWS.

Model-Theories which, though false, are useful

Somehow the ancient peoples survived, even though they believed for their whole, long lives that the earth was flat. When it occurred to them to ask "What's holding us up?" someone suggested it was a man named Atlas, standing on a turtle. It is likely that no one lost any sleep wondering whether Atlas might get tired and drop us or asking "What the dickens is holding the turtle up?" During the Middle Ages, when scholars believed the earth was at the center of the universe which extended as far out as a huge shell studded with stars, no one lost sleep wondering whether our spherical universe might be like a ball hovering in a vast empty space. If they had, they might have wondered whether our fairly large universe was in the center of that infinite, empty space, or perhaps off in one corner, or moving around. I read somewhere that Bertrand Russell, after Einstein died, told a radio audience that the universe appears to be finite, but unbounded. Descartes surely would have had fun with that, not to mention Charles Dodgson and Hans Christian Anderson.

The point is that for 99% of the human race, life goes on exactly the same, whether $2 + 2 = 5$ in some other universe or not. For the remaining 1%, 99% of our lives will go on exactly the same, whether $2 + 2 = 5$ or not. Before agriculture, our ancestors would have felt no need to know precisely when the time for planting corn or maize was at hand. Once agriculture had arrived, they did have that need. It made no difference whether they thought the sun used an underground tunnel to get from sunset-west back to sunrise-east, whether it went down around the earth, or whether the earth goes around the sun. What they saw between the time they woke in the morning and fell asleep at night were the same 'sense impressions.' (Einstein's phrase) Those who grow the crops for us need only to know when to plant. As for us, all we need to know is such simple things as the difference between thumbs and food. (Did we know that back when we were babies?)

Lavoisier (1743-94) and Dalton (1766-1844) both did their chemistry, not sure what was 'behind' the sense-impression appearances, but hoping to find out by sorting through the carefully recorded descriptions of their carefully weighed experiments. Mendeleev (1834-1907) took their hypotheses about unseen and unfelt reality and organized them even farther into today's 'Table of Elements,' and did so still operating on the false idea of solid matter. Thomson (1856-1940) and Millikan (1868-1953) were able to make discoveries of never-seen electrons, still operating under the false Greek uncuttables model. (B. Jaffe's history of chemistry, *Crucibles*, is more exciting than even a Zane Grey western!)

The point of the preceding is the absolute necessity of recognizing two critical truths. First, all human beings can live full human lives, even though they believe much that is false. Even those who know and vehemently disagree about how to interpret modern experiments can live full and fulfilling lives. Secondly, one of the most basic premises of our everyday human lives is this: people who disagree on fundamentals can not both be right. In our view, there are both true and false ingredients in Einstein's and Bohr's opposing but unifying-for-them worldviews. No one is wholly mistaken! In this third millennium, we have the luxury of freely choosing the ingredients which will give us an even more complete and completely true G. U. T. than theirs.

"Straight" and "curved" common-sensically understood

Ask a class of students if "vug" is a word. Ask them whether they would issue a challenge if someone used "vug" during a game of Scrabble. Then experiment. Try to give them the word's meaning by simply repeating it. First, instruct them as follows: "As soon as someone gets the idea, raise your hand." Then repeat "vug" over, and over, and over, and over, and . . .

Now, replace "vug" with "straight." Of course, everyone will instantly know what that means. They will, if they have ever heard of police asking intoxicated drivers to walk a straight line. Or if they recognize the fact that the lines their geometry teachers put on the board are not perfectly straight. They'll probably also know that a straight face is different from a straight line. There are even people who can understand that Hans Christian Anderson wasn't being straight when he referred to non-emperors as emperors in that story about one emperor's new clothes.

One often sees statements such as "a straight line has length but no breadth," and people sometimes wonder how anything can exist which is so lacking in solidity as to have no breadth at all. It is clear of course that such a statement cannot apply to the "straight lines" of the physical world. A line drawn on paper certainly has breadth, and even the boundary between two areas of different colors always has a certain vagueness if we examine it closely enough. In theoretical geometry, a straight line is just something that we think of. In Cartesian geometry, what we think of is really a number. For example, one of the straight lines of the Cartesian plane is $x = 1$; and in the case of this line what it comes to is simply that any number is exactly equal to 1, or not equal to 1 at all. Two numbers are either the same or different. They may be nearly equal, but they do not merge imperceptibly into one another. (E. C. Titchmarsh, *Mathematics for the General Reader*, p.68)

Two millennia before Titchmarsh, Plato — dazzled by the pure thinking done by 'mathematicians' — drew attention to the fact that we can think of paired lines which are far from equal in length, of other pairs closer than the first ones to being equal, and of pairs that are perfectly equal, even though none of us has ever seen a pair of lines perfectly equal in length. (See his *Phaedo*.) It seems as if we have a mental yardstick against which we measure sensed things. Plato (or was it Socrates?) used that fact to take perhaps the first and the greatest stride in the direction of Einstein's later, personal realization that we have concepts which cannot possibly have been derived from sense observation! What would Plato have thought, if he had known that all mathematics, including geometry, is make-believe?

If it is true that poltergeists can throw solid objects into closed rooms, presumably they require a fourth dimension in which to do it. This is a situation with which (as mathematicians) we are perfectly prepared to cope. We have merely to add a fourth co-ordinate, and identify a point with a set of four numbers, say (x, y, z, w). This provides us with a four-dimensional Cartesian geometry. It is impossible to visualize it, but as a mathematical system it is not much more difficult to handle than three-dimensional geometry. In fact we can introduce any number of dimensions in the same way. (E. C. Titchmarsh, *Mathematics for the General Reader*, p.74)

Nick Herbert reports that when Bohr died, he was still thinking about one of Einstein's 'thought experiments.' What is a thought experiment? Nothing less than something little children do so much of: pretending. Let's pretend that light (which no one has ever seen) is tiny billiard-ball photons. We can pretend they move in a straight line, or we can imagine they move in a curved line. Straight the way the crow flies, curved the way a turtle would have to crawl from New York City to Los Angeles. Crows can fly straight (for a short distance) because they can fly through

semi-frictionless air. Turtles cannot fly, hence must travel on the surface
of the earth which, as anyone who looks at a globe knows, is curved. If
billiard-ball photons have no mass that makes them swerve when they
get too near the gravitational pull of the sun, they will travel straight. If
they have mass and are pulled off course by a massive sun, they will
travel in a curve. (Eddington estimated that 160 tons of sunlight fall on
the earth each day.)

> If Lobachevski's geometry is true, the parallax of a very
> distant star will be finite. If Riemann's is true, it will be
> negative. These are the results which seem within the reach of
> experiment, and it is hoped that astronomical observations may
> enable us to decide between the two geometries. But what we
> call a straight line in astronomy is simply the path of a ray of
> light. If, therefore, we were to discover negative parallaxes, or
> to prove that all parallaxes are higher than a certain limit, we
> should have a choice between two conclusions: we could give
> up Euclidean geometry, or modify the laws of optics, and
> suppose that light is not rigorously propagated in a straight line.
> It is needless to add that everyone would look upon this
> solution as the more advantageous. Euclidean geometry,
> therefore, has nothing to fear from fresh experiments. (H.
> Poincare, quoted in A. Eddington, *Space, Time and
> Gravitation: an Outline of the General Relativity Theory*, p.9)

This is a Chinese Box of pretendings. First, we pretend light is the
kind of thing that can travel straight or not-straight. To travel, it must go
from one place to another at a certain speed. That means we must pretend
there are invisible lines called "paths" and invisible notches to mark
where each mile ends and the next one begins. (Path is a negative
concept borrowed from what we experience — or don't experience —
while walking through the woods, namely, the 'area' where there is not as
much grass and weeds as elsewhere.) We pretend there are places, one
for the beginning of the non-existent path, another at the place of each
notch, and another at the place where the non-existent path ends. As for
waves, anyone who thinks a wave can travel in any kind of line has not

Quintalism & Subatoms

done much analysis of, say, water waves which do not exist. At most, zillions of water molecules get squeezed, unsqueezed, resqueezed, etc., and the ones at the surface rise during a squeeze, fall during an unsqueeze, etc. Whoever thinks a rope tied at one end and whipped to make waves is seeing double if they believe the wave is as real as the rope. And so on. The bottom line is: what does "straight" mean when Poincare uses the phrase "rigorously propagated in a straight line"? And, if we rename the straight line "curved," what should we rename a curved line? "Straight"?!

A student wrote a letter to Einstein in 1946 in which she said "I probably would have written ages ago, only I was not aware you were still alive." The student also confessed an inability to understand curved space. Einstein replied as follows:

Dear Master . . . ,

Thank you for your letter of July 10th. I have to apologize to you that I am still among the living. There will be a remedy for this, however.

Be not worried about "curved space." You will understand at a later time that for it this status is the easiest it could possibly have. Used in the right sense the word "curved" has not exactly the same meaning as in everyday language. (H. Dukas & B. Hoffman, ed., *Albert Einstein: the Human Side*, p.108)

Again, what sense does "curved space" now have, since "straight" means "curved." And why would anyone think that space or, worse, space-time, even exists?! An extremely provocative essay was published in the *New York Times* on February 2nd 1999, "Ether Re-Emerges as the *Je Ne Sais Quoi* of Physics." It reported that, in a then recent issue of *Physics Today*, Dr. Frank Wilczek of the Institute for Advanced Study in Princeton, N. J. had argued that, contrary to current thinking, Einstein did not sweep away the concept of the ether. In fact, the ether, "renamed and thinly disguised, dominates the accepted laws of physics." In his own book on relativity, Einstein's thinking shows clearly that his 'field' theory

160

The Wonderful Myth Called Science

is a tweaked version of the same picture older thinkers called "the ether." Chapter XIX begins:

> If we pick up a stone and then let it go, why does it fall to the ground?" The usual answer to this question is: "Because it is attracted by the earth." Modern physics formulates the answer rather differently for the following reason. As a result of the more careful study of electromagnetic phenomena, we have come to regard action at a distance as a process impossible without the intervention of some intermediary medium. If, for instance, a magnet attracts a piece of iron, we cannot be content to regard this as meaning that the magnet acts directly on the iron through the intermediate empty space, but we are constrained to imagine—after the manner of Faraday—that the magnet always calls into being something physically real in the space around it, that something being what we call a "magnetic field.(A. Einstein, *Relativity: the Special and General Theory*, p.63)

Literary critics believe that texts should not be read, first, last, and almost always, with the aim of determining precisely what the writer meant. That, if I remember precisely the verbal text of a literary critic's words to me, is what his words meant. He said something like "If I can make a consistent reading of it that is justifiable, even if my reading is different from someone else's consistent reading or even from the author's meaning." My wife tells me she recalls Hemingway's listener-response to some critic's response to one of his novels or plays, "Gee, I didn't know I meant all of that."

Of course, if we are reading what the Duchess says — or is it Lewis Carroll or perhaps Charles Dodgson? — we can never be certain the author didn't have two or six different meanings in mind, and then the reader who is no longer a naïve child knows the game is to figure out what all of them were. But where people are not playing games for fun (one physics teacher told me that his attitude is that "It's all a game"), we have to assume they are trying to be precise; especially when their appeal is to mathematics and, therefore, mathematical precision.

161

Quintalism & Subatoms

Einstein's mindset is blazingly clear in passages such as the above, and there are many. It marks a return to Descartes' physics which was built on the assumption that there is no vacuum or, in terms used by Aristotle's translators, no void. It rejects Newton's crystal clear thesis that moons and planets move in (near?) frictionless, empty, vacuous space, no part of which has even the slightest trace of difference from any other part, and by his insistence that he would not go 'on record' regarding the nature of the gravity that makes the moon, for instance, keep 'falling' toward earth but never getting there because it inertially keeps tending to move in a straight path, and the two motions add up to one elliptical motion.

A thorough-going relativity theorist must ask "What moves?" Does the moon move? Or does the rest of the universe move in relation to a stationary moon? Does the peanut dropped by a passenger aboard a jet flying from NYC to LA fall straight down to a point on the floor or does it travel ahead in a curve that will allow it to collide with that same point of the forward-moving plane's floor? ("Point" is synonymous with Newton's "relative place," and "peanut" — like "moon" and "earth" — can be interpreted as a "point" in Euclidean geometry.)

Enough, space and time are not infinite!

We use whatever pretending makes predictions easiest.

And everyone has her or his own preferred pretendings or hypotheses or guesses or interpretations. And, once we make up our own mind, we must be careful to sift out the grain from the chaff.

> Let us notice two things about field theory. First, although it assumes that everything we observe can be explained in terms of fields, the fields themselves are never directly observed, perceived by our senses, and so are, at least to some extent, abstractions. That is, fields do not necessarily have to exist. (D. Park, *Contemporary Physics*, NY: Harcourt, 1964, p.124.)

162

The Wonderful Myth Called Science

Fields either exist or they don't, but if they do, it is neither necessarily nor contingently, since those adjectives are projections from logicians' theories about the relations between propositions. (An important point made by atheists who are unpersuaded by Saint Thomas's third proof for God's existence.) In either case, our concepts 'of' fields are abstractions, however useful they are.

When faced with a dilemma of this kind, it is well to go back to foundations. Why do we believe that light is made up of waves? In a beam of light there are no visible crests and troughs to indicate vibrations, as in water. We can measure the relative intensity of light, very simply, and we can measure something we call the velocity of light, but such quantities could be associated quite as easily with a shower of particles. Intensity, for example, could be interpreted as equivalent to the number of particles crossing a unit area in the path of the beam in unit time; just as we do with a jet of smoke. The real reason why we favor the wave hypothesis for light is that a wave is the ideal mathematical model for representing what happens when two rays of light cross, or when light is made to form certain patterns by being reflected from a sheet of polished metal which has been ruled with very fine lines, set close together (i.e., a diffraction grating). It is then found that the wave hypothesis predicts the patterns perfectly, and our belief in the vibratory nature of light is strengthened when we find that we can produce similar interference patterns by undoubted ripples in a tank of liquid. These measurements also yield consistent values for what we choose to call the wavelength (or frequency) of the light. Thus the evidence for the wave nature of light turns out to be no more than the statement that the mathematical theory of waves gives correct predictions when applied to visible light or any other form of radiation.

The same is true of particles. ...'fluid particle' is a figment of the mathematician's imagination, and so is any other particle in the Newtonian sense of the word. There are problems for

163

which the particle is the appropriate basis, and for which it would be foolish to use any other model. Thus although we could represent, if we wished, the trajectory of a cannon-ball by a Fourier series or integral, that is, by waves, it would be a waste of time to do so, because the problem is solved much more easily on the particle basis. There are no 'wave mechanics' in external ballistics.

The mathematician regards particles and waves simply as means to an end, and he is unlikely to be worried by the impossibility of forming a mental image of something which is simultaneously a wave and a particle. To him particles and waves are as real as points in Euclidean space or the operator i, no more and no less. His concern is mainly that the chosen method of representation gives correct predictions. (O. G. Sutton, *Mathematics in Action*, NY: Harper, 1960, p.113-114.)

This book, *Mathematics in Action,* is my favorite, not just for this gem, but for Sutton's explanation of calculus' foundational image or metaphor. Still, water waves do not exist (noted above), and if non-point particles exist at all, they most certainly do exist in Newton's sense of small body. Sutton's passage assumes naïve realism. The predictions are actually predictions of sensations. All the references to cannon balls, light particles, etc., are hopeful as if pretendings that may turn out to be true. But maybe not.

What if we make a mistake?

In 1988, Stephen Hawking offered this suggestion vis-à-vis making mistakes. That was about sixteen years before he recently practiced what he preached by admitting to an error related, among other things, to using black holes to travel to other universes. His suggestion? Don't do what Quine said is always possible.

What should you do when you find you have made a mistake like that? Some people never admit that they are wrong and continue to find new, and often mutually inconsistent, arguments to support their case — as Eddington did in

164

opposing black hole theory. Others claim to have never really supported the incorrect view in the first place or, if they did, it was only to show that it was inconsistent. It seems to me much better and less confusing if you admit in print that you were wrong. (S. J. Hawking, *A Brief History of Time*, Ch. Nine)

When the error is fundamental to one's entire worldview, courage, but especially honesty, is needed to deal with it. Einstein's honesty was as admirable as Hawking's. As noted earlier, Einstein wanted a picture-plus-theory of reality as it is in itself, irrespective of us and our thoughts. But . . .

As Einstein's life drew to a close, doubts about his vision arose in his mind. 'The theory of relativity and the quantum theory . . . seem little adapted to fusion into one unified theory,' he remarked in 1940. He wrote to Born, probably in 1949, 'Our respective hobby-horses have irretrievably run off in different directions. . . . Even I cannot adhere to [mine] with absolute confidence'. In the early 1950s, he once said to me that he was not sure whether differential geometry was to be the framework for further progress, but if it was then he believed he was on the right track. To his dear friend Besso he wrote in 1954, 'I consider it quite possible that physics cannot be based on the field concept, i.e., on continuous structures. In that case, nothing remains of my entire castle in the air, gravitation theory included, [and of] the rest of modern physics'. I doubt whether any physicist can be found who would not agree that this judgment is unreasonably harsh. In one of the last of the many introductions Einstein wrote for books by others, he said: 'My efforts to complete the general theory of relativity . . . are in part due to the conjecture that a sensible general relativistic [classical] field theory might perhaps provide the key to a more complete quantum theory. This is a modest hope, but certainly not a conviction.' (A. Pais, *'Subtle is the Lord...' The Science and the Life of Albert Einstein*, Oxford: Oxford U. P., 1982, p.467.)

Quintalism & Subatoms

It may seem that whatever errors Einstein and Bohr made about the physical world are largely irrelevant to everyday life. But that seeming is an illusion. Life may go on as it has, but what is it that is going on? Any of us who cannot answer that question may be shirking one of our most pressing responsibilities.

Here is just one single illustration — and an easy one at that! — of the relation of modern discoveries (what most people call 'science') to real-life issues. If the only physical bodies that really exist are subatoms, the logical conclusion is that money, thought of as something distinct from protons, electrons, and especially photons, is a figment of our unwary imaginations.

And the concept of money is inextricably woven into any serious thinking about today's 'global economy.' Consider, then, this opinion from someone who has studied the history of the money fiction.

> The dollar is dying; so too are the yen, the mark, and the other national currencies of the modern world. Our global money system is infected with a deadly virus, and, already severely weakened, it is now only a matter of time before it succumbs. The dollar, mark, and yen will join the ducat, cowrie shells, and the guinea in the scrap box of history, as items of interest primarily to antiquarians and eccentrics. (J. Weatherford, *The History of Money*, p.xi)

Dollars are not dying. That is poetry. Dollars have never existed. Thales's question, "What is it really?", will always bring honest people to the realization that — even common-sensically speaking! — a piece of gold is a piece of gold, regardless of whether we call it a dollar, a pound, a mark, or anything else. A rose by any other name. Take the 'dollar bill' into the best FBI laboratory in the country. Nothing but shaped ink marks and possibly fingerprints will be found on it. No wonder some naïve-realist 'philosophers' admit in print that money exists only because we all believe it does. Except we don't all believe it. But if only subatoms exist, not even gold or paper does. So what does?

166

The Wonderful Myth Called Science

> Most dollars today have no physical existence. Even ledger books began to disappear in the 1960's and '70s, replaced by computer records. The Federal Reserve System processes an average of $2.1 trillion each day through an electronic network known as Fedwire. (T. J. Stiles, "As Good As Gold?", in *Smithsonian,* September 2000, p.116)

Now, ask what it is that really goes through "an electronic network." And what really transpires when we 'pay' for something using our cell phone and a credit card 'number'? We can use *Hydrogen: the Essential Element*, John Rigden's 2002 presentation of 20th century theories about what Gerhard Herzberg called "the most important constituent of the universe" to ponder the question and find an answer. For instance, Rigden's sixteenth chapter asks, "What is an electron?" several times. His answer? "As experimental methods become more and more refined and are capable of producing more and more accurate data, the world seems to get curiouser and curiouser." And what becomes especially curious is the very nature of humans and human life.

> As we look to the future, the hydrogen atom will continue to help us meet the challenge of embracing the natural world with understanding and, in the process, to understand better the place of humankind within the larger scheme of things. (J. Rigden, *Hydrogen: the Essential Element*, p.170)

To 'understand that better' demands that we take Adam Smith and Karl Marx and twentieth-century history seriously enough to try and understand how to think of a world without money. It will take an Ockhamist Bulldozer to be thorough here with our 'reality check.' But it can be done, once we realize that, even in everyday-thinking terms, the exchange of goods and services began as barter and never became anything other than barter. The 'monetary system' is nothing but bookkeeping regarding who did what or gave what to whom, who deserves what was done for or given to them, and so on. Or should be only that!

Quintalism & Subatoms

An equally important realization about money is this. The history of money reveals nothing, if it does not reveal the fact that, so far as our conscious experience is concerned, money's just use requires at least as much trust in our companion humans as it does in God. And honesty in trying to learn the truth about our place in 'the larger scheme of things.'

The wonderful myth called "science" has made a change in 'the times.'

Despite ignorance about relativity theory and quantum physics, life for some may continue to go on as before. But for more and more people even now, life does not. And the kind of technology made possible by modern discoveries, if combined with superstition and lack of education, will eventually put all of us in danger.

> Sam Harris presents major religious systems like Judaism, Christianity and Islam as forms of socially sanctioned lunacy, their fundamental tenets and rituals irrational, archaic and, important when it comes to matters of humanity's long-term survival, mutually incompatible. A doctoral candidate in neuroscience at the University of California, Los Angeles, Harris writes what a sizable number of us think, but few are willing to say in contemporary America: "We have names for people who have many beliefs for which there is no rational justification. When their beliefs are extremely common, we call them 'religious'; otherwise, they are likely to be called 'mad,' 'psychotic' or 'delusional.'" . . .

> In the 21st century, Harris says, when swords have been beaten into megaton bombs, the persistence of ancient, blood-washed theisms that emphasize their singular righteousness and their superiority over competing faiths poses a genuine threat to the future of humanity, if not the biosphere. . . (N. Angier, "Against Toleration," in *New York Times Book Review,* 9-5-04, p.19)

Such charges are serious. To ignore them becomes less and less excusable for those who regard themselves as responsible citizens of a single world where all of us must live together, and where honest discussion of our own 'facts' and our own 'reasons' is the only alternative to violence and war.

Transition. Logical consistency

The next and final chapter will deal with a scandalously ignored contradiction between current claims about 'science' and one of the most incontrovertible conclusions that follow logically from modern discoveries.

What contradiction? On the one hand, the most vocal advocates of the wonderful myth called "science" insist that it is the product of sound logical reasoning tested by observation. On the other hand, logical reasoning tested by observation shows that practically every conclusion regarded as a scientific fact by the most vocal advocates of 'science' is based on a naïve concept of observation, one proven false by an overwhelming mass of evidence generated by the very modern discoveries thought to be based on it.

Since practically all of the observing relied on by today's experts is the observing done while reading, it will be easy for each of us to do our own logical reasoning about observation, and then to test it by and on the first-hand observing we do while reading about observing.

CHAPTER V.

I See Colors

There is no such thing as a conceptual definition of this distinction (aside from circular definitions, i.e., of such as make a hidden use of the object to be defined). Nor can it be maintained that at the base of this distinction there is a type of evidence, such as underlies, for example, the distinction between red and blue. Yet, one needs this distinction in order to be able to overcome solipsism.

A. Einstein, *Reply to Criticisms.*

A PROLOGUE. TRUE SCIENCE DEMANDS LOGICAL CONSISTENCY

Undrawn conclusions.

Maybe I should have known better. I already had in my possession all the pieces of a solution to that puzzle at least. It simply never occurred to me to put two and two, or fifty and fifty, together. I'm referring to what I wrote — or quoted — at the end of a four–hundred page doctoral dissertation on H. H. Price's theory of sense-perception.

> We do not put forth these "starting-points for an alternative solution," nevertheless, without a measure of trepidation. Each of them has been challenged, and many present day philosophers will regard at least some of them as having been definitively refuted. On the other hand, there is the consolation of knowing that many theories, at one time regarded as definitively refuted, are later seen to have contained their share of truth after all. There is also the

consolation of realizing that the actual situation in contemporary philosophy justifies the serious attempt to rethink the problem, despite the risks involved, since the truth—as one writer has well expressed it—is that "the problem of perception remains the most unresolved in the whole of epistemology."

The citation, "the problem of perception remains the most unresolved in the whole of epistemology," came from page one hundred twenty–two of Kenneth Gallagher's *Philosophy of Human Knowledge*. He used two chapters of his book to discuss the 'extremely tangled skein of puzzles' bunched together as 'the problem of perception.' Unable to untangle the puzzles, he declared the problem unresolved.

Recently, it struck me that the problem of perception had been solved. To see that fact requires a simple shift in perspective. Instead of looking out over the spread of opinions and noticing only their incompatibility, we can take Sherlock's view, verify that all the bases have been covered, eliminate the false opinions, and find the truth that has been right before our very eyes the whole time!

Contrary to what Gallagher and I were thinking, the problem of perception was not unresolved. It had been solved numerous times by 1967. It had been solved in every way possible by 1967. Each 'new' post-1967 solution has been just one more instance of old wine being poured into new skins.

Perhaps I should have known that in 1967. To compose a critical presentation of Price's version of the sense-datum theory, I myself had studied about fifty different solutions. They covered all the essentials. One key to understanding that fact is Einstein's attack on the fallacy that we extract our concepts from sense experience. I was 90% in agreement with that, since I had surrendered naïve realism four years earlier. Had it occurred to me to ask, "Can there be any new solutions essentially different from those now available?" I'm not really certain what I'd have answered. But it didn't occur to me, and I didn't draw the logical conclusion from what I already knew.

The Wonderful Myth Called Science

In retrospect, my failure partially fits a point made by M. Cohen in a 1961 printing of *A Preface to Logic*, a book in which he shows how thoroughly our everyday thinking is saturated with the 'rules' of logic, but often un-applied.

> The notion that deductive reasoning must necessarily be a sterile series of tautologies arises from the failure to distinguish between psychological, physical, and logical considerations. Psychologically it is obviously not true that the conclusion is always contained in the premises. For ages men accepted the elementary laws of arithmetic without seeing that they involve as a necessary consequence the proposition that there are no two numbers whose ratio is the square root of two. Or, to take a more concrete example, I may know that the *Camperdown* was sunk and none aboard could be saved, and I may know also that Smith sailed aboard that ill-fated vessel. And yet it may be some time before the union of these two propositions flashes on my mind the startling conclusion that Smith must have been drowned. (M. Cohen, p.26)

Undrawn conclusions or contradictions?

There is still another crucially important lesson here. Logical consistency is what older books about logic focused on. They also took note of the opposite of consistency, namely, inconsistency or contradiction. If I knew that the ship sank, that every person aboard it had drowned, and that Smith was one of those aboard, I would have contradicted myself if I then denied that Smith was drowned.

That is what I did when I agreed with Gallagher that the problem of perception was unresolved. Not only hadn't I unified everything I knew in order to draw the valid and true conclusion. I contradicted that valid and true conclusion.

Contradictions are much easier to spot in what others profess to believe than in what we ourselves think. We all tend to argue negatively. When we cannot give reasons why we are right, we can at least tell why others are wrong. Even when we are wrong, we tend to defend our errors by focusing on what we regard as the even worse errors of others.

And others do make errors, and we can spot them once we become critical thinkers. That is the negative facet of Sherlock Holmes's scientific method. But not everyone is ready to be a critical thinker. For instance, I often test students to see if they can detect the error in this passage from a 1963 book on economic realities.

> Money is not real. What made it seem real for so long was its scarcity. Since money is supposed to be spent on things, its scarcity can truly reflect reality only when that reality is made up of a general scarcity of things. It no longer is, except mostly by intention. Since the imagined scarcity of money no longer mirrors an existing scarcity of things, and since money that is not meaningfully scarce is not exactly money, the whole nature of the symbol has changed profoundly. (D. T. Bazelon, *The Paper Economy*, pp.13-14)

Bazelon's book is immensely helpful in breaking down for the uninitiated the vast number of complex pragmatic fictions invented by experts to somehow get an instant glimpse of billions of daily exchanges, by billions of individuals, of endlessly varied goods and services. It is helpful in the same way that literary-criticism books are helpful when we try to get a glimpse of the infinitely varied contents of a modern library.

But how can money be scarce *if it is not real*? What kind of money isn't 'exactly money'? For instance, last week's news reported that the world's richest nations have agreed to cancel or forgive forty billion dollars of debt owed by poor nations. Where is all of that owed money now? Where will it go? Will it just evaporate? And will 'the dollar' remain stable long enough for the relief to amount to a full forty billion of them?

Or consider a reply made by Carl Rogers to Richard Evans who was interviewing him in the 1970's.

The Wonderful Myth Called Science

Rogers: . . . For me, the perception is reality as far as the individual is concerned. I don't even know whether there is an objective reality. Probably there is, but none of us ever really knows that. All we know is what we perceive, and we try to test that in various ways. If it seems to be perceived in the same way from several different aspects, we regard it as real. The world of reality for the individual is his own field of perception, with the meanings he has attached to those various aspects. Probably any organism, certainly the human organism, is always trying to satisfy its needs as they are experienced in the phenomenological field—that is, the world about him as he perceives it, in the reality as he perceives it.

Evans: You appear to be agreeing with Immanuel Kant who suggested that there is no reality except in terms of man's perception of it.

Rogers: Yes, I am. I have tried stating that sometimes and find that it always leads to fruitless arguments, so I don't say so very often. But, as you suggest, it really fits in with my response to your earlier comment. None of us knows for sure what constitutes objective reality and we live our whole lives in the reality as perceived. (R. I. Evans, *Carl Rogers: the Man and His Ideas,* pp.9-10)

"None of us knows for sure what constitutes objective reality." Did Rogers ever wonder whether or not Evans really existed and whether he, Rogers, knew what Evans was thinking?

Inconsistencies can co-exist with vitally important truths.

We must pick and choose our truths from a mix of truth and error. For instance, what Rogers was thinking expresses one of the most important lessons we need for dealing with others. In a sense, every one of us lives in our own 'world' or 'reality.' But that means that everyone else lives in their own 'world,' too. "Being sensitive to others" means "being able to see the 'world' the way the other person sees it" or "putting ourselves in the other's moccasins." It's one of the great lessons of the novel and

175

I See Colors

movie *To Kill a Mockingbird.* Rogers's insight was the inspiration for the one-on-one therapy he invented and called "client-centered therapy."

Together with mine and Gallagher's case, those two examples of failing to draw the logical conclusions from one's beliefs, even contradicting oneself, illustrate another obvious truth. No one is entirely wrong, and many who are wrong in some fundamental ways can help us learn other valuable truths. A major reason why many debates, e.g., that between Einstein and Bohr, go on and on is because each party grasps an important truth and refuses to surrender it.

That last point is extremely important. For instance, Noam Chomsky was only half right with the chief argument he used to blow B. F. Skinner's language theory out of the water. Skinner claimed that when people speak or write, they are repeating what they have heard. Chomsky pointed to the constant output of verbal behavior that is obviously novel or new. That part of the Chomsky torpedo relates to Einstein's thesis about thought, not language, being creative. How, for instance, did Skinner acquire his new concept of 'operant conditioning,' a necessary addition to balance off Pavlovian, stimulus-response conditioning? Where did he get his ideas of positive vs. negative reinforcement, of extinction, and so on? Certainly not from observing them. He was explicit. All we observe are the pigeons, rats, or other organisms behaving vis-à-vis ping-pong balls, mazes, pellets, meat, words, etc.

But, from a naïve-realist point of view, Skinner was right to point out that, apart from reinforcement (and imitation!), there is no immediately obvious reason why children learn to not say "go'ed" as the past of "go," the same way we use "toed" as the past of "toe." Why "went"? Why "ate"? Why "thought"? (Does anyone seriously believe that irregular-verb use is related to innate neural networks?)

But how would Skinner explain the continued outpouring of novel novels? And the output of novel Ph. D. dissertations? And the constant proliferation of novel journals in every 'field' that are needed for the novel publications required for tenure and promotion? And the indefinite (not infinite) possibilities for the creation of novel sites on the Internet?

176

The Wonderful Myth Called Science

If language existed, Chomsky's exposé might have sufficed. But Skinner could still ask whether there is any better answer than his to the question, "Why do infants raised in Boston grow up as speakers of English, infants raised in Paris grow up as French speakers, infants in Cairo as Arabic speakers, etc.?" Does every language have matching irregular verbs? What are the French, Arabic, Chinese, etc., counterparts for the past of "go," "eat," "think," etc.?

Incidentally, in the same debate between Skinner and Rogers referred to in Chapter IV, Rogers turned Skinner's behaviorism against him in what logicians would call a *reductio ad absurdum*. But Quine was right. If someone is determined not to change their opinion, they make some adjustment. Skinner's honesty was at times astonishing. He accepted Rogers's accusation that his [Skinner's] principles logically led to a conclusion which is absurd. But only to the rest of us! Here is what Rogers said, just prior to the statement by Skinner which was quoted earlier.

> ROGERS: I said, "From what I understood Dr. Skinner to say, it is his understanding that though he might have thought he chose to come to this meeting, might have thought he had a purpose in giving this speech, such thoughts are really illusory. He actually made certain marks on paper and emitted certain sounds here simply because his genetic makeup and his past environment had operantly conditioned his behavior in such a way that it was rewarding to make these sounds, and that he as a person doesn't enter into this. In fact, if I got his thinking correctly from his strictly scientific point of view, he as a person perhaps doesn't exist." I thought I would draw him out on the subjective side of why he was there, but to my amazement he said he wouldn't go into the question of whether he had any choice in the matter, and added "I do accept your characterization of my own presence here." I've wondered ever since. I would like you to comment further on that, because I think it gets close to the heart of some of the difference between us.

I See Colors

SKINNER: I think it does. I think it's very close. There is this strange feeling that if you deny the individual freedom or deny an interpretation of the individual based upon freedom and personal responsibility that somehow or other the individual vanishes. This is not at all the conclusion one could arrive at. I think you can make the assumption that each person is completely determined to do what he is now doing and is going to do by his own genetic and environmental history. I say this to my class . . . . (In *Carl Rogers: Dialogues,* ed., H. Kirschenbaum & V. L. Henderson, pp.93-94)

How prescient Rogers was became clear in 1983 when Skinner published the third volume of his autobiography.

I am willing to concede that I have committed a kind of intellectual suicide in writing this autobiography. When Anthony Trollope confessed that he had written his novels partly for money, his reputation suffered. By tracing what I have done to my environmental history rather than assigning it to a mysterious, creative process, I have relinquished all chances of being called a great thinker. . . . If I am right about human behavior, I have written an autobiography of a nonperson. I have collected alms for oblivion, but not, I think, for no 'reason.' There are consequences. (B. F. Skinner, *A Matter of Consequences*, pp.412-13)

In the context of everyday thinking, for anyone who wishes to gain some self-knowledge and to become aware of the extent to which his or her thought habits appear to be 'shaped' by society or culture, Skinner's writings, especially *Beyond Freedom and Dignity*, are among the most valuable of the library's resources. In those first years after we have begun to form ideas in response to the speech of the significant others in our lives, our rapid acquisition of ideas and attitudes seems almost entirely the product of our social environment. Watson, Hitler, and Mussolini knew that. Parents who fight to have additions to Darwin taught in the schools and parents who fight to keep them out know that. The authorities who censor what 'the public' gets to hear and read know that. When Socrates preached the need for living the examined life, what he wanted Athenians to examine were not other people's cultures, but their own.

Reflective theist believers will understand why atheist believers see 'religion' as something dangerous, when the history of religious wars, crusades, persecutions, executions, etc., is reviewed. Of course, reflective atheist believers who are honest admit that the atheist regimes of Hitler and Stalin more than evened history's score.

The challenge now is this. Can we visit the library, read extensively, pick the gems of wisdom from the accompanying errors, and assemble enough of them to reach the goal Einstein sought, a true grand unifying theory?

Solving the perception-problem: experimenting by —while — reading.

Suppose I write "You can see" and "You cannot see." If you can understand what you just read, two contradictory thoughts came to you. If you can reason logically, you can be confident that, if one of those contradictory thoughts is false, the other must be true. If you can see, you will know that "You cannot see" must be false. By eliminating that false half of the contradiction, you will know that the other half, "You can see," is true.

This one experiment can test dozens of long-disputed questions. For instance, it can test the validity of Descartes' revolutionary thesis that concepts do not originate from sensation. (Did you ever see your self? Or seeing?) Kant and, later, Einstein made Descartes' thesis one of the basic beliefs in their respective belief-systems.

It can also test twentieth century debates about meaning and, by doing so, verify the *Intelligo*, the principle that "everybody understands everybody else's meanings very well indeed." (Rorty)

And thirdly, it can test debates about reference, which are intimately related to debates about meaning. "Reference" is a noun extracted from the verb "referring," as in "I'm referring to you!" Like debates about meaning, debates about reference are utterly crucial in relation to this chapter's clinching argument regarding the myth called science and the scientific method. To understand what is meant here, in this chapter, by "I see color," it is essential to know what "I" refers to, which of all my doings "see" refers to, and above all what "color (as such!)" refers to.

The importance of reference.

The reason for beginning this final chapter by referring to a switch from believing in 1967 that the problem of perception was unresolved to just the opposite now in 2005 is this. Apart from the mythical distinction of 'scientific' knowledge from everyday thinking, etc., the greatest obstacle to achieving a unifying theory is not the incompatibility of Einstein's field with Bohr's quantum, but unclear ideas about sense perception. Einstein and Bohr, like those who are on opposite sides of debates about human evolution, appealed to logic, parsimony, and observation as the deciding factors in what they freely willed to believe.

In fact, the problem of perception is not literally a problem of perception. It is a problem of thinking. Perceiving, observing, or seeing are not difficult. Thinking with utter precision about selves, their conscious acts, and especially the proper object of the act of seeing is extraordinarily difficult. Unless . . .

Unless we get as clear as possible about what we mean when, for instance, one of us says something as innocuous as "I see a couple of brown spots on this banana." And unless, like a bulldog with its piece of meat, we maintain our bite on what we mean while reading about all of the issues that relate to it and that invariably make us lose our way from time to time. There is no fear that we can lose our way so badly that we never regain it. Regardless of our abstract theory about seeing what we see, we can always instantly revert to our everyday thinking. In fact, we cannot help doing so.

Here are a few more thoughts about reference, that is, about knowing exactly what real thing, real thought, or object of a real thought we are referring to at any given moment.

180

The Wonderful Myth Called Science

A child who believes in Santa Claus may picture a male human with a long white beard when he or she thinks of Santa. If the child is taught to believe in a god, the child may picture a male human with a long white beard when he or she thinks of that god. How can the child know those two names refer to two different beings?

We already know that Ayer spoke for many when he said that "God does not exist" is nonsense or has no sense, meaning by "sense" what is also meant by "meaning." Suppose we reject Ayer's nonsensical principle. Now "Santa exists," like "God exists," has a meaning. But can its meaning be different from the meaning of "God exists"? We might say that the meanings can only be different if the child uses those statements to refer to two different beings. What if only one exists? What if neither exists? What about the child who believes both exist? How can we ever know what any thoughts or words refer to unless we also know how many and what kinds of things there are to refer to?

A quick review of the library's holdings will show that gallons of ink have been used to answer such questions once and for all. But there is no once and for all. Like each of us, each writer was able to have his or her own 'last word,' and their last words diverge radically. Quine, referred to earlier, gave his own controversial answer in *The Roots of Reference*, published in 1974. It clarifies or attempts to clarify the extraordinary claim he made in his 1950 "Two Dogmas of Empiricism," a claim which is all tangled up in issues about referentiality—or extension —or supposition. (Many different names have been used for the issue.)

> As an empiricist I continue to think of the conceptual scheme of science as a tool, ultimately, for predicting future experience in the light of past experience. Physical objects are conceptually imported into the situation as convenient intermediaries—not by definition in terms of experience, but simply as irreducible posits comparable, epistemologically, to the gods of Homer. For my part I do, qua lay physicist, believe in physical objects and not in Homer's gods; and I consider it a scientific error to believe otherwise. But in point of epistemological footing the physical objects and the gods differ

181

only in degree and not in kind. Both sorts of entities enter our conception only as cultural posits. The myth of physical objects is epistemologically superior to most in that it has proved more efficacious than other myths as a device for working a manageable structure into the flux of experience. (W. V. O. Quine, *From a Logical Point of View*, p.44)

From this 1950 passage, it seems logical to draw the conclusion that, for Quine, the only thing anyone can refer to is past experience, future experience, plain experience, and the flux of experience. It also seems logical to ask just how different that conclusion is from Ayer's nonsense about "God doesn't exist," nonsense of which that all of us can make perfectly good sense. Quine was logical (!) perhaps, in making that claim if his goal was to stay above the fray in which naturalist-materialists believe Homer's mention of gods cannot refer to any gods, since there are none to refer to, the fray in which they are opposed by theists who insist that when they talk about a god they are referring to a real being who is over and beyond, more than, in addition to, or besides, anything and everything physical or material.

An aside. In his 1980 *Philosophy & the Young Child*, G. Matthews offers a revealing footnote to the Quine biography. It comes in Matthews' chapter on Piaget and in connection with the distinction between conceptual schema and external reality. At one point in critiquing Piaget's view, Matthews is reminded . . .

As I read these words, there beat in on me the words of my teacher W. V. Quine, spoken to me when I was a graduate student. We were discussing whether the element of what philosophers call "intentionality," that hallmark of the "inner" and the nonphysical world, could be eliminated from reports of what someone is thinking. I was skeptical, Quine insistent. "Let's face it, Matthews," he said earnestly, "it's one world and it's a physical world." (G. T. Matthews, *Philosophy & the Young Child*, p.47)

182

The Wonderful Myth Called Science

The reading experiment will therefore rely on what has been posited previously, namely, that the first question for all who pray to be truly wise must be "What exists?" Quine's reference to references to gods and bodies seems to refer also to culture, and if to culture, then to societies that embody them. He also refers to concepts, situations, intermediaries, definitions, posits, errors, entities, myths, and structure. The question is, "Do any of those last things exist?"

That question cannot be answered unless we know what "those last things" refers to. In fact, questions of meaning and reference go hand-in-hand. That is our hypothesis. It goes with the hypothesis that each of our answers to any question must have, as its mindset-context, an overall philosophy about everything (else) that exists.

A quick review of the mindset-context for a reading-observing experiment .

The negative half of reaching the truth is getting rid of the errors. That is why so much attention has been given to reductionism, Ockham's Razor, etc. Part of the positive half of reaching the truth is recognizing the absolute need we have for various models and pretending-thoughts or — in terms more and more commonly used — for constructs, pragmatic theories, etc., even when we know that those pragmatic fictions are false, which is why we call them fictions. Unless we unmask them as errors, albeit indispensable as false pretendings, they will block our ascent or assent to the pinnacle of insight.

The positive side is the quintalist answer to "What exists?" It is an alternative to over-specied common sense and under-specied dualisms and monisms. There are five species of things that exist. Not fifty thousand or two or one. Five. Unless there are only four, and I am mistaken about the fifth. In that case, the four are persons, understood thoughts, experienced sense data and images, but no bodies, not even subatom-sized ones.

Equipped with that overall framework of things (in general), we will have adequate room in our model for theories about the things that exist and those that don't, plus thoughts about what things do, and why they do what they do. The information on all five counts just waiting in the library for serious readers is staggeringly large.

My confidence comes from thoughts such as the following. Einstein was right to think that knowledge of reality improves. The worldview we today can construct is, in fact, wholly reliable in essentials. Future research will not overturn any of the basic principles maintained here. What is known by experts about the so-called 'laws of nature' may be improved by further details, but that is all.

Those are generalizations. The particulars are as follows. There is no fear that we will someday discover that the earth really is flat, as our ancestors believed. There will never be a return from today's table of elements to Aristotle's air—earth—fire—and—water foursome. So long as we can destroy cities with a single bomb, no one will go back to fighting with swords and spears. And, given modern word processors and other communications media, no one can expect the deluge of ever new books, magazines, newspapers, etc., to dwindle to the trickle that the famous complainer named Qoheleth was moaning about more than two thousand years ago: "Of the making of many books there is no end." (Ecclesiastes 12:12) He didn't know what "many" meant!

That knowledge is available to those who can read. Each learner must do his or her own reading. A clear understanding of what goes on while we read will make it possible to improve the non-bodily, psychological side of Descartes' worldview. Much of that upper half of Descartes' revolutionary worldview has already been discussed in the preceding chapters, even in the previous chapter that seemed to focus on the lower, material part of his worldview. Without learners, the world which Einstein insisted is 'out there today, irrespective of our observation' would still be there, but as unknown by any human persons as it was at the time of the alleged Big Bang. It may seem silly to go in circles that way, but there is a reason for it. Those who make reality dependent on us and our knowledge of it have to play fast and loose with words in order to dodge the obvious objection, "Wasn't the world here before you?" Like, did your parents just come into existence the moment you first opened your eyes and looked at them?! Ah, yes, and did you observe your eyes before you opened them? Consistency!

Why I didn't realize that the problem of perception had been (re)solved.

A large part of the reason was old, self-defeating thought-habits. The worst of them was the learned-partitioning of what I had learned by the time I was thirty. Or maybe the worst was the fact that the foundational commonsense learning I had done between my birth date and my fifth or sixth birthday wasn't even on my radar screen. It counted for nothing. It was entirely pre-scientific. It was for illiterate aborigines and naïve six-year-olds, and what do they know?! As for serious, higher learning, mine was divided into empirical science (nothing perennial), perennial wisdom (St. Thomas's), and theology (St. Thomas's). All three had their own evidence sources, their own methods of using those sources, and their respective conclusions.

By 1967, the walls had come tumbling down, but the project of reworking so much accumulated knowledge(s) that had been kept separate rather than integrated was going to take many more years to complete. Final unification consists in creating a master framework that is both pared down to essentials but not pared down one step too far.

The third reason is that in 1967 I just didn't know enough. I still had a long way to go in learning more fully the important insights which great thinkers, from Bacon to Husserl, had achieved. In 1967, the added years of studying those thinkers in order to teach their ideas in class were still in the future. So, too, were the ten-plus years spent in another graduate program, learning the theories of the great philosopher-psychologists, from Freud and Jung to Allport and Rogers. Most of all, I had not yet made the full acquaintance of William James, far and away the greatest of modern thinkers.

The plan for this concluding chapter.

It is now Saturday evening, June 25, and I am convinced this whole project can be completed by focusing on the three major concepts

I See Colors

abstracted from the chapter title, "I see colors." First is my concept of my self, the agent who does the seeing. Second is my concept of the act of seeing or being aware. Third is my concept of color or colors. Agent-act-object. I see colors thus parallels I understand thoughts. My thoughts are theory, and my theory must explain both myself seeing colors and myself understanding thoughts. Three items:

1. My self.

2. My act of 'seeing.'

3. The colors I see.'

The most difficult problem any of us must deal with is making sure we fully grasp what we believe is the right theory about each item and making sure each theory is 100% logically consistent with the other two. None of the three theories can even be fully grasped unless all three are grasped together in a kind of magic moment of dawning insight. Consistency!

Three procedural notes.

First, there will be some elimination by humor. Also known as the *reductio ad absurdum* or the unmasking of what everyday thinkers recognize as absurdity. Regard the references above to parents and eyes as examples.

Martin Gardner used two quotations as epigraphs for his 1981 *Science: Good, Bad, and Bogus*. The second quote can be used to test his title. It comes from the American journalist, Henry Louis Mencken (1880-1956). H. L. M. wrote "One horse-laugh is worth ten thousand syllogisms." Did Gardner, who shocked the world, or at least me, with what he revealed in *The Whys of a Philosophical Scrivener*, fall into the same type of self-refutation that Martin Luther King said Saint Thomas used when he wrote that an unjust law is no law? If true, every law would have to be just, since there cannot be any such thing as an unjust one. Can bogus science be science? Doesn't all science have to be genuine in order to be science? And what can bad science be? Good-sense mind shifts can solve such apparent contradictions.

The Wonderful Myth Called Science

Second, the movie-going experience is indispensable. We buy our ticket. We hand it to the usher who tears it in half and gives back one of the new wholes which is only half the size of the earlier whole, we take our seat, the lights dim, the picture suddenly appears on the screen at the same time that sounds blare from the speakers, and we begin an integrated movie-experience which can be theorized-about or thought-about in three radically unsame ways. Suppose the movie is Hitchcock's *Vertigo*. We can slip effortlessly into thinking about three individuals, two men and one woman, and a many-weeks (months?) scenario telescoped into a two-hour abridgement. Or we can mentally 'step back' for a critical assessment — a naïve-realist one — of what is really going on, namely, that we are sensing a stream of visual and auditory sensations produced by unsensed causes, namely, the film projector and sound system, and those sensations evoke associated memory-images correlated with ongoing thought to produce the illusion that we are witnessing real events in real people's lives.

To learn what is really, really going on, however, it is necessary to go one step further, the way we take Eddington's two tables further. What is really going on does not involve a theater, movie screen, or loudspeakers. Those things do not exist, only swarming, dancing subatoms. They are 'out there,' whereas our sense-data, memory-images, and thoughts are 'in here.'

Thirdly, there is the process of generalizing inference. A major component of our commonsense philosophy is that (i) there are other beings essentially like us and that (ii) their experience is essentially like ours. Thus, even though I have no idea who you are who sees what you are seeing this very instant, and now this instant, and now this one, and now . . . I am confident that there are essential similarities between you (if you exist) and me (not you), and between what is going on as you (not I) see this and what goes on when I (not you) read. I have relied on that unprovable assumption in the past, and it has always 'worked.'

That is, during my naïve-realist period, it never occurred to me that I might be the only person in existence. Later, however, it did occur to me. But I am still confident that others existed and do exist. I am still

187

confident that I've been able to understand what Einstein had in mind when he wrote, what Ayer had in mind when he wrote, and so on. And, during class, my experience is that what I sense provides clues that I correctly interpret as 'I am in class,' 'this student is (not) understanding me,' etc. I assume the same will be true of you, whoever you are.

Of course, we are all different, in both similar and different ways. But variety is not inconsistent with logical consistency.

Summary.

Next to taking account of all the relevant evidence, logical consistency is critical for acquiring a grand unifying theory or philosophy.

B. MY SELF

Introduction. Who is "I"?

Who is I? Quine talked about an I. Who was he referring to? (Reference is crucial here.) Whoever or whatever that I was, he believed that it continues to think, it does believe, and it considers. Einstein referred to I, too. Einstein's I is convinced, chooses, apologizes, exists among the living, can't adhere, considers, and doubts. Matthews refers to I as well. In addition to the things Quine and Einstein said it does, it is also skeptical. We (you and I) know that I is currently writing this book's fifth chapter describing a reading-observing experiment so that, if you ever come across I, you can check to see if I also reads.

But perhaps you have already met I? Is it perhaps you? For instance, would you say "I am writing this"? The I I know rather doubts it, since it knows that you aren't writing it, which you would have to be doing in order to say truthfully, "I am writing this."

What is truly amazing is that young children, even before they go to school, know who I is. You can test them with games about it. Ask "What are you thinking right now?" When they answer "I am thinking . . . ," reply "No, I don't want you to tell me what I am thinking; tell me

The Wonderful Myth Called Science

what you are thinking." And it's clear from their listener-response behavior that they know that you (who's that?) are toying with them.

The reason for what precedes, which might look like a bit of persiflage, is simple. Not until a learner fully appreciates the profound mystery involved in each reader's ability to know which of the universe's billions or zillion zillions of entities is being referred to (who's reading and who's writing?), will that learner learn to appreciate the gigantic step forward in the history of Western thought Descartes took when he so skillfully composed what appears to be a diary entry written solely for his own eyes. (You can do it, i.e., read that.) It was nothing less than an invitation to us readers to learn a revolutionary new way to think about genuinely scientific knowledge of reality. His revolutionary new way marks a transition from (i) traditional if-then, hypothetical-at-best, thinking to (ii) existential-foundational science.

From if-then generalizing to thinking about real-life existents.

Asked to raise their hands if they have ever taken a course in which they learned something about the brain, students readily comply. The sole reason why any of them hesitate is because they're uncertain how much the 'something' had to be for it to count. They take it as a simple joke when asked the follow-up question. "Whose brain was it? Yours? The teacher's? Mine? Whose brain is 'the' brain?!"

And yet, that simple, effortless shift, instantly recognizable by every college student, constitutes the very heart and soul of Descartes' great revolution. For his predecessors, scientific knowledge was general knowledge, universal knowledge. For many today, scientific knowledge about 'the brain' is the same thing, viz., a lot of generalizations, usually formulated as 'laws,' about all brains. (If any exist.) Although Descartes may not have fully noticed what he was doing, he was switching from a search for certainty about theories which 'abstract' from concretely real things, to a search for certainty about actually-existing, concretely real things.

Let's pretend that traditional Western science began one day when Thales looked around and once more 'saw' what everyone else saw,

189

namely, numerous individual things. For instance, up there were some drifting clouds against a blue sky. Yesterday, when the whole sky was dark, raindrops by the millions fell from up there to the earth. Some fell into the rivers and lakes, some fell and made puddles that soon dried up. Last winter, in northern regions, snowflakes by the millions fell instead of raindrops, and when it was cold enough the water near the surface of ponds and lakes froze into ice. But today was today, and all that past was now just memory. Today, sitting on the side of a hill, he could see a town in the distance, feel the earth beneath him, look up at the clouds, and just think.

Then it struck him. What if he tried to look beyond just this or that cloud, this or that raindrop, river, lake, etc. He had volumes of memory about the changes he had seen in nature. Perhaps all those changes were interrelated. Perhaps there was a reason why clouds turned to rain, rain turned to ponds, ponds froze into ice, and ice melted back again into water that could be boiled to form steam which rose to form clouds, etc., etc. Perhaps there is some invisible 'stuff' that all of those things are 'made of.' Perhaps what he looked at and felt were just apparently different. If he could learn about that stuff and how it worked, he wouldn't have to examine each cloud individually, each drop of rain, each snowflake, etc., to know true reality.

If we think of the 'stuff' as matter, then we can think of forms as that which shapes homogeneous matter into different things. Taught by Socrates, Plato applied the idea that there is a distinction between sensing and conceiving. Though we sense those many individual things, we can think of them with generalizing ideas. When we think of many virtuous people, we do so with a generic concept of virtue. We think of many just laws with a generic notion of justice. We think of many beautiful things with a generic idea of beauty. In time, the idea came to Plato that, in addition to the many virtuous people, just laws, and beautiful things, there is something more, namely, eternal, unchanging, purely intelligible Forms or Ideas of Virtue, Justice, Beauty, etc.

Aristotle, too, began with the many individual things. He also agreed that truly scientific knowledge about clouds, rain, and other things in

190

nature is final and unchanging. No need to rewrite the books on physics every few years. His solution was to view the changeable things in nature as combinations of Thales' matter and Plato's forms (after bringing them back down to earth!). True knowledge consists in mentally extracting things' forms from their matter. Once extracted from changing material things, the form-ideas are frozen in our mind, and we can study their relations to other frozen, extracted form-ideas.

Aristotle's approach to geometry, briefly described in the preceding chapter, helps to understand his point of view. Geometers would not call their knowledge "scientific" if they knew everything about just one, concrete, roughly-drawn triangle. Scientific geometry consists in the knowledge of abstract principles that allow the geometer to know ahead of time that, if he or she ever encounters another triangle, it too will have three sides, the sides will meet to form three angles, and the sum of the angles, regardless of the length of the triangle's sides, will always add up to 180^0. Once possessed, that knowledge will remain forever true, even if some triangle-hater destroys every real triangle — or pyramid — that exists.

The result of both Plato's and Aristotle's views of unchanging scientific knowledge was to divert attention from really existing, always changing, individual things. Attention was directed at unchanging, abstract, general ideas and principles used to understand how we can think of many things at once. (Chapter II)

Descartes, though he absorbed the best of the new physics and discovered a new, superior form of geometry useful for understanding it, had an attack of good old-fashioned common sense. He began the break with science-is-knowing-general-principles-of-what-might-exist view as opposed to science-is-knowing-what-here-and-now-does-exist view. The debate between Bellarmine and Galileo was not about abstract physics. It was not about mathematics. It was about reality. The real planets, the real earth, the real sun.

And, when Descartes set about trying to explain which here-and-now, really-existing things he could be certain of, it occurred to him that the first really-existing thing he could be certain of was his own real self. *His own real self.*

Descartes used Sherlock's method. First, elimination. When we turn to his *Meditations*, we see that his first question is, "Can we — no, can I! — be certain that bodies exist?" His answer was, "No, not if I rely solely on my senses." This became Kant's and Einstein's view, and the reasoning of all three thinkers begins with the new physics and physiology summarized in a brief passage in this book's previous chapter.

We come into this world belief-less. We — at least our bodies — are bombarded immediately by heterogeneous stimuli — light, sound, odors, heat, etc., coming from the environment where they are jumbled together. First, they are filtered by our five sense organs. Eyes pick out light, ears filter out sound, skin responds to kinetic energy, resistance, and so on. The different or heterogeneous stimuli, after being filtered, trigger non-differentiated or homogeneous electro-chemical chain-reactions conveniently lumped together under "nerve-impulses." Only these homogeneous nerve impulses, traveling via afferent neurons, reach the brain, at which point . . .

Next, he asked, "Is mathematics certain knowledge?" No. First, we are all prone to error in our calculations. Besides, what if some all-powerful deceiver has put a permanent hex on our thinking, so that we always err in our calculations? Even more importantly, it makes no difference whether our actual calculations are flawless or error-ridden. By themselves they say nothing about what exists outside our mind.

That is how Descartes eliminated uncertainties in order to zero in on the real things he decided he could be absolutely right about. He was certain he could understand the thoughts he was thinking, whether they were true or false. He could understand that, if wax exists, it can change its appearances and remain wax. But what if wax doesn't exist? It may be that $2 + 2 = 4$ in every conceivable world, but what if only one thing, I, exist? Thoughts are not certainly true unless they match certainly-existing realities. But "I, Descartes, exist" must be true. How else could he be understanding all these thoughts? Even to think false thoughts, he had to be real.

Consistency. Something told Descartes that "I know I don't exist" is as self-contradictory as "This thought is false," "This circle is square, that is, not circular," "I am thinking about absolutely everything without exception, but I know there's more than what I'm thinking about," "I am writing this book, I and you are one, but you are not writing it," etc. (He even tells later on what he thinks is telling him those things.)

Each of us must take our own turn testing, "I'm thinking, so I must exist."

Have you been reading? Have you been understanding any of what you have been reading? Do you remember anything you've read? Do you exist? Test yourself. Use Sherlock's method. Think of alternatives. If you can understand them, you can also understand that you have choices. Here are your first two.

I am understanding this vs. I'm not understanding a word of this

I do exist vs. I do not exist

Descartes was a pioneer. He did intimate that it was possible to begin the construction of a new system by doubting everything, that is, by not assuming anything at all. Subsequently, critics pointed out that, in order to be certain that he was thinking and therefore had to be existing, he had to first have the right concept of thinking and the right concept of existing, even had to assume that there couldn't be thinking without a thinker. So he wasn't beginning from absolute zero.

His critics were right, of course, which made his best reply both right and wrong. (It reinforces the idea that real, here-and-now existents, not abstract possibilities, were his quarry.)

Principle X.
That conceptions which are perfectly simple and clear of themselves are obscured by the definitions of the Schools, and

that they are not to be numbered amongst those capable of being acquired by study [but are inborn in us].

> . . . And when I stated that this proposition I think therefore I am is the first and most certain which presents itself to those who philosophise in orderly fashion, I did not for all that deny that we must first of all know what is knowledge, what is existence, and what is certainty, and that in order to think we must be, and such like; but because these are notions of the simplest possible kind, which of themselves give us no knowledge of anything that exists, I did not think them worthy of being put into words. (R. Descartes, *Principles of Philosophy*, Part One, trans., E. Haldane and G. Ross)

He was right to believe that everyone able to read could understand, at least vaguely, what "I am thinking" meant and who the "I" in question was. (Don't you?) And he was right because, without naming it, he used everyday thinking (Einstein) or his common-sense philosophy (here), especially its 'existential content.' It is the indispensable foundation for all higher learning, whether that touted by the Greeks or that sought by himself.

From common sense to quintalism: the first carry-over commitment in regard to I.

The Greeks, if they existed, never doubted that nature, i.e., stars, sun, moon, and earth, exists. They did not doubt that we could learn about it, which is why they debated about nature's nature and about the relation of our knowledge to it. True, Parmenides the mystic proclaimed that all things are one and unchanging, but his aberration was quickly co-opted by the brilliant reasoner, Plato, for his dualist framework that had ample room for a world of sensed particulars, including human bodies, a host of individual psyches, each able to be incarnated in individual bodies, as well as an unsensed world of mystic realities perceivable only by the intellectual part of humans' psyches.

In fact, the Greeks generally would have been surprised by the question, "Why does everyone think the world is made up of stars, sun, moon, etc., and keep arguing about it in those terms?" The reason is simple. There's only one world that we all live in and wonder about, and it's the world with stars, sun, moon, etc. The medieval schoolmen whom Descartes mocked had the same basic orientation.

After Descartes, confidence in such everyday thinking was severely jolted. In less than a century after Descartes' death, Berkeley denied that any physical world exists. Hume argued that the quest for certainty about it is eternally doomed to fail. Kant thought he had found the key to renewed hope that we can understand reality. But then Kant's project hit a snag. 'The' self. Or was it his self?

The psyche was Kant's snag. He wound up with a profound contradiction at the very point where his two worldviews — of virtual 'reality,' only apparently outside, vs. reality in itself, really outside — intersected. Kant made a sharp distinction between his everyday-thinking idea of his self and the real nature of the? his? self. The difference is very easy to understand for anyone who has ever thought about reincarnation. Probably all of us grow up thinking we are boy or girl, very smart or only average, living in this country and not that, with these parents and not those, etc. But if, in an earlier lifetime, we grew up thinking we were a different sex, with a different IQ, living in a different country, born of different parents, etc., then the question is: "What is my permanent self like, once it is stripped of those non-permanent differences?"

Kant went further. According to physics, we seem to be made up of parts, but we can't prove that. According to physics, every thing we do is governed by iron-clad laws, but we can't prove that. According to physics, we are a temporary part of a temporal physical world, but we can't prove that. On the other hand, we can't prove we are a single thing with no parts, can't prove we have a free will, and can't prove we are immortal. It's as if unprovable science, which says we aren't one, free, or immortal, is at odds with our unprovable sense that we are one, free, and possibly immortal. We can be sure only of how we seem, but not of how we really are. But, Kant, says, we can take a leap of faith and believe whatever seems most reasonable. However . . .

I See Colors

However, which self do we (we?) . . . Which self do I think will take my leap of faith? The one that science says isn't free to make choices? Or the one I can't be sure is real enough to make any choices? If the first, it will only be an apparently free act of faith. If I think — make an act of faith — that it's the self which I'm not sure even exists, the one that I am or am not free to think is free to make the choice, then that will make two acts of faith. Made by whom? Based on what? No wonder a great thinker threw up his hands and declared that Kant can't be straightened out. He's one big mess!

The train of thought that wends its way from Descartes, through Locke, Berkeley, and finally Hume, can lead us into a Kantian labyrinth from which there is no escape. That is why, once we have discovered the true importance of our early common-sense philosophy, the worldview (a theory, a set of beliefs) that serves as the springboard for all this later, allegedly higher learning, we can fill in that huge gap in Descartes' final worldview. We simply add "Our bare ideas, whether of possible knowing, possible existing, possible certainty, etc., are already in place when we begin the examined life." That way, we can at least retain his crucial, scientific discovery that our concepts do not originate in sense experience.

The way to avoid Kant's black hole is to not get close enough to be sucked in. Confronted with the choice, "I am certain that the thought, 'I am,' is true" versus "I'm quite at a loss to know whether 'I exist' is true or not," the learner who has managed to avoid becoming confused will know which is the truth. And, even if that learner does succumb to doubt while poring over the books in the library, the sense of self will return as soon as meal-time arrives, announced by hunger pangs felt by a self for whose reality there can be no lasting doubt.

At any moment, you can test yourself. Are you real? Are you able to understand the difference between "I exist" and "No, I don't"? Do you feel as if you are under compulsion to believe you don't? Or do you feel that, if you try long and hard enough, you can believe "I exist" is true? Or perhaps you don't exist because you are I writing this. Perhaps there is nothing being read or written.

The Wonderful Myth Called Science

Try to imagine your self as a solipsist

Solipsism is the belief that you are the only being in the universe. A solipsist is someone who believes nothing and no one else exists but him or herself. The English word, if reversed, comes from the Latin for "self, alone." Are you perhaps the only being in the universe?

I have looked through my own handful of books by and about Einstein. The only one which lists "solipsism" in its index is the Schilpp volume, *Albert Einstein: Philosopher-Scientist*, which contains the passage cited at the beginning of this chapter. The only other reference in that volume's index is to "Einstein's Influence on Contemporary Philosophy" by Andrew P. Ushenko. Ushenko argued that it is impossible to adopt Einstein's relativity science and to be a solipsist. In other words, one who accepts Einstein's relativity must believe that there are other people in the universe besides him or herself.

The facts are exactly the reverse. In the early 1940's, Einstein was asked to write an essay to be included in an anthology devoted to the philosophy of Bertrand Russell. Einstein took the occasion to praise Russell for a remarkable passage from the latter's introduction to his 1940 *An Inquiry Into Meaning and Truth*. Here is the passage Einstein praised.

> We all start from "naive realism," i.e., the doctrine that things are what they seem. We think grass is green, that stones are hard, and that snow is cold. But physics assures us that the greenness of grass, the hardness of stones, and the coldness of snow are not the greenness, hardness, and coldness that we know in our experience, but something very different. The observer, when he seems to himself to be observing a stone, is really, if physics is to be believed, observing the effects of the stone upon himself. Thus science seems to be at war with itself; when it most means to be objective, it finds itself plunged into subjectivity against its will. Naive realism leads to physics, and physics, if true, shows that naive realism is

false. Therefore naive realism, if true, is false; therefore it is false. (*An Inquiry Into Meaning and Truth*, Introduction.)

There are two key points in Russell's summation. The first is ". . . observing the effects of the stone upon himself." Descartes concluded that bodies cause effects in us, specifically in our brain. One's own brain, if it exists, stands forever between one's self and the rest of the physical universe. That is one of the most important of Descartes' discoveries. It is the inescapable implication of all current neuroscience. That crucial fact must be highlighted and branded on our mind: *Descartes, if he existed, discovered that your brain, if it exists, stands forever between you and the rest of the physical universe.*

Whatever we experience is, is caused by, or is correlated with those effects produced in us. Locke, Berkeley, and Hume took that conclusion as something amply proven by modern science, which of course they called philosophy. Hume, the one who woke Kant from his dogmatic slumber, was the most blunt of the three. In a passage that preceded Eddington's two-tables lesson from modern physics by almost two centuries, Hume repeated in 1748 what Berkeley had discovered:

> It seems evident, that men are carried, by a natural instinct or prepossession, to repose faith in their senses; and that, without any reasoning, or even almost before the use of reason, we always suppose an external universe, which depends not on our perception, but would exist, though we and every sensible creature were absent or annihilated. Even the animal creation are governed by a like opinion, and preserve this belief in external objects, in all their thoughts, designs, and actions.

> It seems also evident, that, when men follow this blind and powerful instinct of nature, they always suppose the very images, presented by the sense, to be the external objects, and never entertain any suspicion that the one are nothing but representations of the other. This very table, which we see white, and which we feel hard, is believed to exist, independent of our perception, and to be something external to our mind,

which perceives it. Our presence bestows not being on it; our absence does not annihilate it. It preserves its existence uniform and entire, independent of the situation of intelligent beings, who perceive or contemplate it.

But this universal and primary opinion of all men is soon destroyed by the slightest philosophy, which teaches us, that nothing can ever be present to the mind but an image or perception, and that the senses are only the inlets, through which these images are conveyed, without being able to produce any immediate intercourse between the mind and the object. The table, which we see, seems to diminish, as we remove farther from it; but the real table, which exists independent of us, suffers no alteration: it was, therefore, nothing but its image, which was present to the mind. These are the obvious dictates of reason; and no man, who reflects, ever doubted, that the existences, which we consider, when we say, this house and that tree, are nothing but perceptions in the mind, and fleeting copies or representations of other existences, which remain uniform and independent.

So far, then, are we necessitated by reasoning to contradict or depart from the primary instincts of nature, and to embrace a new system with regard to the evidence of our senses. But here philosophy finds herself extremely embarrassed, when she would justify this new system, and obviate the cavils and objections of the skeptics. (D. Hume, *Enquiry Concerning Human Understanding*, Sec. XII, Pt. I)

"But this universal and primary opinion of all men is soon destroyed by the slightest philosophy." By the slightest science, he would have written in 1940. Thus science, according to Hume, undermines any hope of creating a sound science and exposes us to all the doubts of the skeptics.

This relates to Russell's second key point. "Thus science seems to be at war with itself; when it most means to be objective, it finds itself plunged into subjectivity against its will." Russell was far more consistent and clear on this than Einstein, which makes Russell's writings extremely valuable for understanding the real truth about objectivity and subjectivity. Nothing of Russell's is better for grasping — with no punches pulled — what it means to say that 'naïve realism is a plebeian illusion' than his 1927 work entitled variously as *Philosophy* and *An Outline of Philosophy*. That, its unambiguous 'spelling out' what the 1940 passage meant or ought to have meant, is the work's chief value. For instance:

> ... what we know indubitably through perception is not the movements of matter, but certain events in ourselves which are connected, in a manner not quite invariable, with the movements of matter. To be specific, when Dr. Watson watches rats in mazes, what he knows, apart from difficult inferences, are certain events in himself. The behavior of rats can only be inferred by the help of physics, and is by no means to be accepted as something accurately knowable by direct observation. . . To return to the physiologist observing another man's brain: what the physiologist sees is by no means identical with what happens in the brain he is observing. . . In a strict sense, then, he cannot observe anything in the other brain, but only the percepts which he himself has when he is suitably related to that brain. (B. Russell, *An Outline of Philosophy*, 1927, pp.140, 147)

To the views of Descartes, Locke, Berkeley, Hume, Kant, Einstein, Eddington, and Russell, we can add the following arguments from authority.

> Descartes' line of argument is perfectly clear. He starts with reflex action in man, from the unquestionable fact that, in ourselves, co-ordinate, purposive, actions may take place, without the intervention of consciousness or volition, or even contrary to the latter. As actions of a certain degree of complexity are brought about by mere mechanism, why may not actions of still greater complexity be the result of a more

The Wonderful Myth Called Science

refined mechanism? What proof is there that brutes are other than a superior race of marionettes, which eat without pleasure, cry without pain, desire nothing, know nothing, and only simulate intelligence as a bee simulates a mathematician? . . .

It must be premised that it is wholly impossible absolutely to prove the presence or absence of consciousness in anything but one's own brain, though, by analogy, we are justified in assuming its existence in other men. (T. Huxley, "*Animal Automatism,*" 1874, pp. 196-97)

We are accustomed to talk of the "external world," of the "reality" outside us. We speak of individual objects having an existence independent of our own. The store of past sense-impressions, our thoughts and memories, although most probably they have beside their psychical element a close correspondence with some physical change or impress in the brain, are yet spoken of as inside ourselves. On the other hand, although if a sensory nerve be divided anywhere short of the brain we lose the corresponding class of sense-impressions, we yet speak of many sense-impressions, such as form and texture, as existing outside ourselves. How close then can we actually get to this supposed world outside ourselves? Just as near as but no nearer than the brain terminals of the sensory nerves. We are like the clerk in the central telephone exchange who cannot get nearer to his customers than his end of the telephone wires. We are indeed worse off than the clerk, for to carry out the analogy properly we must suppose him never to have been outside the telephone exchange, never to have seen a customer or any one like a customer—in short, never, except through a telephone wire, to have come in contact with the outside universe. (K. Pearson, *The Grammar of Science*, 1892, pp.60-61)

We are such thoroughly visual animals that we hardly realize what a complicated business seeing is. Objects are 'out there', and we think we 'see' them out there. But I suspect that

201

really our percept is an elaborate computer model in the brain, constructed on the basis of information coming from out there, but transformed in the head into a form in which that information can be *used*. (R. Dawkins, *The Blind Watchmaker*. p. 34.)

First, a fundamental principle: The brain exists in order to provide an internal representation of "reality." Quotation marks are employed here in deference to the fact that no creature, including ourselves, can ever know any other "reality" than the representations made by his brain. (R. Restak, *Brainscapes,* 1995, pp.3-4)

Plato said that we are trapped inside a cave and know the world only through the shadows it casts on the wall. The skull is our cave, and mental representations are the shadows. The information in an internal representation is all that we can know about the world. (S. Pinker, *How the Mind Works*, 1997, p.84)

For me, this is an easy question. I believe that animals have feelings and other states of consciousness, but neither I nor anyone else has been able to prove it. We can't even prove that other people are conscious, much less other animals. In the case of other people, though, we at least can have a little confidence since all people have brains with the same basic configurations. (J. LeDoux, quoted in "God (or Not), Physics and, of course, Love: Scientists Take a Leap," *New York Times,* 1-4-05, p. D-3)

The last three are prominent neuroscientists, experts on research into brains, if brains exist. Every last scrap of recent 'scientific research' fits the conclusion which each of the three, if they exist(ed), stated with perfect clarity and distinctness. (The only qualification needed is the addition of "directly" to Restak's statement that no one can know any reality other than their own representations. If our conceptual theories are true, Einstein insisted, then they can be the means of indirectly knowing external reality.)

Einstein's relativity theories, if true, add to the solipsist challenge.

The pillar on which the entire edifice of Einstein's relativity rests is the finite velocity of light, if light exists. It is the postulate that nothing can move from one place to another faster than light can. And light cannot do it instantaneously. The true 'paradoxes' of relativity theory are not length-shortening or time-dilation. The true paradox is that, if we cannot see anything till light sparks rods and cones to send impulses via afferent neurons to our brain, we cannot experience anything farther away than the effects produced in our brain. Since no light reaches our brain, the darkest place in town, how is it that any would-be scientists can fail to realize what Russell meant by saying that none of us experiences anything but effects in us? How can anyone fail to understand Einstein's praise for Russell's succinct summation of the argument against naïve realism?

Why does anyone believe merely-inferred light exists? Or takes time to travel?

An experiment that brings to the surface your inescapable self-recognition.

Have you been reading what you see here? Or writing it? Perhaps you're not certain. If not, test yourself. Were you standing on your head while you read or wrote the last two pages? Were you running your eyes or your hand across the lines of print? If you close your eyes, can you still feel the seat of the chair beneath you? If the chair you are sitting in suddenly disintegrated, do you believe you would feel a change before you hit the floor? Are seeing and feeling the same sense? Are seeing and hearing the same sense? Can you see your hearing and feeling? Can you hear your seeing and feeling? Can you see your seeing? Can you hear your hearing?

Those questions were designed to partially misdirect your attention. Here now is the question related to this section of this final chapter.

I See Colors

While you read those questions, did you keep wondering who was being referred to by "you"? Or were you concentrating more on the doings referred to, such as standing on your head, reading, running your hand across the lines, feeling, seeing, and hearing? Now that your attention is being directed to "Who is 'you'?", do you think you know the answer? Martin Gardner tells of a student who asked "How do I know that I exist?" The professor's answer? "Who's asking?" If you're not sure who is the 'you' referred to in the previous paragraph, ask yourself "Who's not sure?" (Chapter 1 in Gardner's *The Whys of a Philosophical Scrivener* may be the best introduction to solipsism ever written.)

The key point. Normally, we are aware of vastly more than we pay attention to or notice. I know what I've been thinking about for the past hour, I know I have not been at the beach or standing on my head during that time. And I did not spend the hour repeating over and over, "Don't forget what you're writing about, be ready to say you haven't been standing on your head or at the beach, and above all keep in mind all the things you're now reminding yourself not to forget!" This 'awareness' that accompanies our mature seeing, hearing, feeling, and thinking about other things is so easy to overlook, yet crucial for retaining the essential convictions of our everyday thinking. This little-noticed awareness is like a constant 'sense' flavoring one's own personal, private, subjective stream of consciousness.

It is here, in this 'background awareness,' that we will find the evidence for our self, the agent who can do all the things we think of when we speak of standing on our head, reading, etc. As long as puzzling questions are not being raised, we refer naturally and easily to our self. The expressions used by the writers quoted above exhibit this tacit feature of everyday thinking. Each unself—consciously uses first-person pronouns. LeDoux gives the most accurate report of all. He did not avoid one I's commitment by hiding behind a fuzzy we.

. . . the greenness, hardness, and coldness that we know in our experience

. . . This very table, which we see white, and which we feel hard

The Wonderful Myth Called Science

. . . what we know most indubitably through perception

. . . we are justified in assuming its existence

. . . We are like the clerk in the central

. . . no creature, including ourselves, can ever know

. . . all that we can know about

. . . I believe that animals have feeling

The individual-personal context of all would-be 'science.'

A diverse group of twentieth century thinkers known as "existentialists" stressed the fact that 'the sciences' — physics, chemistry, biology, etc. — must be seen in their true context, individual humans' individual lives. The best presentation of this context is Albert Camus's classic description of the possibly futile quest for eternal certainty. He gave dramatic force to the presentation by starting his essay with what he called "the only genuine 'philosophical' problem."

> There is but one truly serious philosophical problem, and that is suicide. Judging whether life is or is not worth living amounts to answering the fundamental question of philosophy. All the rest—whether or not the world has three dimensions, whether the mind has nine or twelve categories—comes afterwards. These are games; one must first answer. And if it is true, as Nietzsche claims, that a philosopher, to deserve our respect, must preach by example, you can appreciate the importance of that reply, for it will precede the definitive act. These are facts the heart can feel; yet they call for careful study before they become clear to the intellect. (A. Camus, *The Myth of Sisyphus,* trans. .J. O'Brien, p.3)

Camus's question, "Is life worth living?" is quintessentially personal. Suicide (till recently) is one person's act of ending one person's life, and we know which person it is. Otherwise we'd normally call it homicide. Camus's whole life and the slow development of his thinking is worthy of study. The newspaper essays he wrote from 1944-1947 regarding

I See Colors

French national policies during the transition from occupied country to restored unification are particularly instructive for our troubled political times. (See *Between Hell and Reason*.) But it was in *The Myth of Sisyphus*, that he most dramatically put the quest for a grand unified theory into its real-life, personal context.

> Whatever may be the plays on words and the acrobatics of logic, to understand is, above all, to unify. The mind's deepest desire, even in its most elaborate operations, parallels man's unconscious feeling in the face of his universe: it is an insistence upon familiarity, an appetite for clarity. . . . Yet all the knowledge on earth will give me nothing to assure me that this world is mine. You describe it to me and you teach me to classify it. You enumerate its laws and in my thirst for knowledge I admit that they are true. You take apart its mechanism and my hope increases. At the final stage you teach me that this wondrous and multicolored universe can be reduced to the atom and that the atom itself can be reduced to the electron. All this is good and I wait for you to continue. But you tell me of an invisible planetary system in which electrons gravitate around a nucleus. You explain this world to me with an image. I realize then that you have been reduced to poetry: I shall never know. Have I the time to become indignant? You have already changed theories. So that science that was to teach me everything ends up in a hypothesis, that lucidity founders in metaphor, that uncertainty is resolved in a work of art. What need have I of so many efforts? The soft lines of these hills and the hand of evening on this troubled heart teach me much more. (Pp. 19-20)

A thread runs directly from Camus to the passage from Russell's *Inquiry Into Meaning and Truth* which Einstein singled out for comment. Once again, here is the crucial part of that passage:

> The observer, when he seems to himself to be observing a stone, is really, if physics is to be believed, observing the effects of the stone upon himself. Thus science seems to be at

war with itself; when it most means to be objective, it finds itself plunged into subjectivity against its will.

". . . it [science] finds itself plunged into subjectivity against its will." Sheer poetry! Science has no will. This exercise in personification says more dramatically what "the observer observes effects upon himself" states more prosaically. Anything that counts as science is thought, and every real thought, whether it is about atoms or suicide, is understood by a real person and is therefore as subjective as the sensory effects the thoughts are about. At any moment, you can check — verify or falsify — Russell's claim. Ask whether (i) you are experiencing any thoughts that are not your own and (ii) whether you are experiencing any sense effects other than those which are in you. Which 'you'? The one whom you call "I."

The subjectivity of all 'objective' science

If anyone assents to the thought that there are other thinkers and other thoughts, the assented-to thought is subjective to the person who does the assenting. And if there are, indeed, other thinkers and other thoughts, then — vis-à-vis the assentor — they are part of objective reality as fully as anything else that exists, whether subatoms, energy, masses, moons, tomatoes, or pens. From a god's-eye-point-of-view, humans' thoughts are both subjective and objective the way every proton is both large and small and the way every photon is both at rest and moving. Those are all relative terms. In every case, accuracy demands the qualification "relative to whom or what?!" All three are crucial for Einstein's relativity and Bohr's quantum.

What about a proton? If it actually exists, it must have definite attributes. Otherwise, there would be no difference between it, a photon, or an atom. It must be large relative to a photon, small relative to the area occupied by the entire 'atom.' (That is true only if never-seen photons, protons, and atoms all exist.) The idea is a carry-over from everyday thinking in which we know that the man in the street is huge compared to the ant crawling on the sidewalk next to the Empire State Building, but tiny compared to the Building. It makes no sense whatever to say something is just large, period!, or just small, period!

As for rest and motion, anyone who understands Newton's generalization of Galileo's relativity knows that we can pretend the photon is at rest and then pretend to use it to anchor the imaginary coordinate system we'll use for mathematically describing what every other subatom in the universe is doing at one of those given instants that Laplace referred to in his famous dictum and that all physics research assumes.[*] Whether Mach or Einstein was right can be tested only by selecting real subatoms to anchor imaginary 'coordinate systems.'

Finally, it is essential to avoid the universally confused and confusing uses of the words, "objective" and "subjective." Like other words, they are ambiguous. They can be used as synonyms for "true" and "false," as in "You would do well to be a little less subjective, that is, to take a more objective view of the matter." Some use "subjective" pejoratively to mean "opinionated," but "opinionated" can have an honest, "I have a view that may be false, but I'll try to be objective, non-dogmatic, and unbiased" sense.

Russell's use of "subjectivity" is different. What follows will be stipulative definitions of "subjective" and "objective" as used in the present context. Actually all honest answers to "How do you define that?" are stipulative; the key issue is "Who is doing the stipulating?" They will be stipulative in the way that "faith" has been given a critically important stipulative definition for this book. These stipulative definitions of "subjective" and "objective" relate to solipsism.

----

[*] If a super intellect knew for an instant all about every atom, it could predict the future down to the last moment in time.

The Wonderful Myth Called Science

Here then are the stipulative definitions of "subjective" and "objective." Since each of us is a knowing subject, and since everyone else, if any exist, is an object of our knowing, all of one's own beliefs are subjective vis-à-vis oneself. My free choices of what to believe depend upon me. If I go poof, my beliefs go poof as well. That is what Bertrand Russell was getting at with his claim that "science is plunged into subjectivity against its will." My 'scientific' knowledge will always be subjective to me, because it will always be my knowledge, depend on me, and depend upon how I interpret the sensory effects private to me. Other persons' knowledges, like those persons themselves, are independent of me, do not depend on me. Since they are objects of my knowledge (objects assented to by my faith), they and their knowledges are objective relative to me. But their 'scientific' knowledge is subjective to them.

Other threads that must now be tied together are these. After Olaus Roemer in the 1670's had studied the eclipses of Jupiter's moons for a considerable length of time, he discovered that the obvious way to explain their regular irregularities was to postulate that the light reflected from them takes a finite amount of time to reach earth. It takes longer when Jupiter and earth are farther apart, less time when they are closer. But Roemer never saw light. Only its effects.

Can you see in the dark? If not, it is because no one ever 'sees' until light reaches his or her own retinas, located at the back of his or her pair of eyes. How can anyone professing to be 'scientific' claim to see anything distant from their own eyes? Other people are always distant and always too large to get inside one's eyes. What observers observe is not grass, stones, snow, rats, brains, or other people, only "effects" produced on or in the non-other observer's own eyes, brain, or . . .

Inescapable subjectivity

The idea of the inescapable subjectivity of experience frightens people. Or at least is unwelcome. The image of oneself being shut off from the outside world is also a picture suggesting we are locked up inside ourselves, unable to learn about the outside world.

I See Colors

What about Einstein? There is reason to think Einstein was not as clear about sense perception as the Russell he praised. Here again is Russell's passage, but with some parts of Einstein's commentary added.

[The] more aristocratic illusion concerning the unlimited penetrative power of thought has as its counterpart the more plebeian illusion of naïve realism, according to which things "are" as they are perceived by us through our senses. This illusion dominates the daily life of men and of animals; it is also the point of departure in all of the sciences, especially of the natural sciences.

These two illusions cannot be overcome independently. The overcoming of naïve realism has been relatively simple. In his introduction to his volume, *An Inquiry Into Meaning and Truth*, Russell has characterized this process in a marvelously concise fashion:

"We all start from 'naïve realism,' i.e., the doctrine that things are what they seem. We think grass is green, that stones are hard, and that snow is cold. But physics assures us that the greenness of grass, the hardness of stones, and the coldness of snow are not the greenness, hardness, and coldness that we know in our experience, but something very different. The observer, when he seems to himself to be observing a stone, is really, if physics is to be believed, observing the effects of the stone upon himself. Thus science seems to be at war with itself; when it most means to be objective, it finds itself plunged into subjectivity against its will. Naïve realism leads to physics, and physics, if true, shows that naïve realism is false. Therefore naïve realism, if true, is false; therefore it is false." (*An Inquiry Into Meaning and Truth*, Introduction.)

Apart from their masterful formulation these lines say something which had never previously occurred to me. For, superficially considered, the mode of thought in Berkeley and Hume seems to stand in contrast to the mode of thought in the

210

natural sciences. However, Russell's just cited remark uncovers a connection: if Berkeley relies upon the fact that we do not directly grasp the "things" of the external world through our senses, but that only events causally connected with the presence of "things" reach our sense organs, then this is a consideration which gets its persuasive character from our confidence in the physical mode of thought. For, if one doubts the physical mode of thought in even its most general features, there is no necessity to interpolate between the object and the act of vision anything which separates the object from the subject and makes the "existence of the object" problematical. . . .

Hume saw that concepts which we must regard as essential, such as, for example, causal connection, cannot be gained from material given to us by the senses. (A. Einstein, *Ideas and Opinions*, pp.30-32)

The more we investigate the contradictions which Thales and his successors began noticing, the more obvious it becomes that we do not see the grass, stone, or snow at all. We do see. But what we do see are extra things, experienced effects in us, distinct from unexperienced causes outside of us. That is why there is no way we can prove grass, stones, or snow even exist.

Einstein agreed . . . theoretically. But it is one thing to assent to a thought, quite another to 'really feel' what the thought implies. Tolstoy's *Death of Ivan Ilych* is a famous parable designed to make us really feel the difference between a detached assent and really feeling what we are agreeing to. He used the schoolboy syllogism, "All humans are mortal, I am human, therefore someday I'll die, too," to illustrate it. Agreeing that naïve realism is false seems to be a detached admission for many. It's comparable to the assent most of us give to "$E = mc^2$." How about Einstein?

It is really our whole system of guesses which is to be either proved or disproved by experiment. No one of the assumptions can be isolated for separate testing. In the case of the planets moving around the sun it is found that the system of mechanics works splendidly. Nevertheless we can well imagine that another system, based on different assumptions, might work just as well.

Physical concepts are free creations of the human mind, and are not, however it may seem, uniquely determined by the external world. In our endeavor to understand reality we are somewhat like a man trying to understand the mechanism of a closed watch. He sees the face and the moving hands, even hears its ticking, but he has no way of opening the case. If he is ingenious he may form some picture of a mechanism which could be responsible for all the things he observes, but he may never be quite sure his picture is the only one which could explain his observations. He will never be able to compare his picture with the real mechanism and he cannot even imagine the possibility or the meaning of such a comparison. But he certainly believes that, as his knowledge increases, his picture of reality will become simpler and simpler and will explain a wider and wider range of his sensuous impressions. He may also believe in the existence of the ideal limit of knowledge and that it is approached by the human mind. He may call this ideal limit the objective truth. (A. Einstein & L. Infeld, *The Evolution of Physics*, p.31)

"It is really our whole system of guesses which is to be either proved or disproved by experiment." That is, "by observation." The mechanisms and laws of nature are hidden from us and we can only speculate or guess about them. A friend of mine carried a copy of that powerfully-stated passage in his wallet for use in barroom 'dialogues' with students from a nearby polytechnic institute.

Modern discoveries are not for the faint-hearted. It is why we are taken back by Einstein's comment, "The overcoming of naïve realism has been relatively simple"? It isn't! It isn't at all, especially when the

212

The Wonderful Myth Called Science

solipsist specter replaces it. John Horgan, in the epilogue with which he ends *The End of Science*, describes what a solipsist experience feels like.

> Years ago, before I became a science writer, I had what I suppose could be called a mystical experience. A psychiatrist would probably call it a psychotic episode. Whatever. For what it's worth, here is what happened. Objectively, I was lying spread-eagled on a suburban lawn, insensible to my surroundings. Subjectively, I was hurtling through a dazzling, dark limbo toward what I was sure was the ultimate secret of life. Wave after wave of acute astonishment at the miraculousness of existence washed over me. At the same time, I was gripped by an overwhelming solipsism. I became convinced—or rather, I knew—that I was the only conscious being in the universe. There was no future, no past, no present other than what I imagined them to be. I was filled, initially, with a sense of limitless joy and power. Then, abruptly, I became convinced that if I abandoned myself further to this ecstacy, it might consume me. If I alone existed, who could bring me back from oblivion? Who could save me? With this realization my bliss turned into horror; I fled the same revelation I had so eagerly sought. I felt myself falling through a great darkness, and as I fell I dissolved into what seemed to be an infinity of selves. (J. Horgan, *The End of Science*, p.261)

This segment began, you will recall, with Ushenko's idea that Einstein's relativity and solipsism are incompatible. The truth is just the reverse. If seeing depends on light, and if none of us ever experiences or observes anything but effects produced in or by brains set in motion by nerve impulses triggered by light, then each of us should take a second look at the thought-experiment discussed toward the end of Chapter III above, the one which Einstein described in Chapter XX of *Relativity*. We should not picture ourselves in the external world looking into a closed watch. On the contrary, we should picture ourselves in a 'spacious chest' or closed elevator, unable to take a look at whatever is outside, other people included.

I See Colors

Theory alone, that is, conceptual thinking, *is the only way we can ever learn what else exists out beyond the sense effects produced in us.* Einstein's 'elevator' thought-experiment shows the distinction between effects experienced inside and guesses about what is on the outside. His hope was to make objective(ly true) subjective guesses about what is objective.

There is some irony in Einstein's apparent agreement with Russell's claim that, "when it [science] most means to be objective, it finds itself plunged into subjectivity." It seems to be at war with an earlier wish to practically escape himself and the inescapable consequences of his admission that we are like elevator riders unable to see what is going on outside the elevator. Here is a telling passage from his 1918 "Principles of Research," a paper delivered at a celebration of Planck's sixtieth birthday.

> To begin with, I believe with Schopenhauer that one of the strongest motives that leads men to art and science is escape from everyday life with its painful crudity and hopeless dreariness, from the fetters of one's own ever shifting desires. A finely tempered nature longs to escape from personal life into the world of objective perception and thought; this desire may be compared with the townsman's irresistible longing to escape from his noisy, cramped surroundings into the silence of high mountains, where the eye ranges freely through the still, pure air and fondly traces out the restful contours apparently built for eternity. (A. Einstein, *Ideas and Opinions*, p.220)

Whoever fully appreciates the solipsistic implication that some would draw from the postulate that nothing in the universe can travel faster than light can now appreciate why that 'escape' dreamed of by Einstein is impossible. His own relativity was based on a principle that explains why every human's knowledge is inescapably subjective. Einstein never used those terms, but the statement in his and Infeld's *Evolution of Physics* is a remarkably candid summation of his theory that 'we' and 'he' can never peek outside to compare our inside guesses with outside nature.

214

The Wonderful Myth Called Science

But each of us who guess about reality that lies beyond our private sense-experience can be certain of at least one self or subject, the one referred to when we say "I exist." The real challenge is not to be certain that "I exist" is true. The real challenge is to find the true answer to "What kind of being am I?"

Change "The mind-body problem" to "The self problem."

The time has come to catch up with Socrates and Descartes, that is, to reproduce in our own learning what they learned before us. Unless they were both lying, they both came to believe in human immortality. Both began young by believing in human bodies, complete with arms, legs, torso, shoulders, head, and brain. But both came to the conclusion that humans can exist and enjoy conscious experience without a human body. They did so, not by virtue of religious faith, but by sound reasoning. To catch up with them must be done individually, by each of us who undertake the examined life.

As explained in Chapter III, one test for open-mindedness is to look at your hand and compare your naïve-realist view of it to the Rutherford view. A follow-up test is to ask, "What would it be like to exist without a body?", and to then weigh the evidence for this answer: "You are doing it now and have always been doing it." That is the only coherent answer for anyone who accepts the modern discoveries made by Descartes, Locke, Berkeley, Dalton, Helmholtz, Thomson, and especially Rutherford. To accept their discoveries and then to believe that there are human bodies — or animal bodies, for that matter — is logically inconsistent.

The best way to begin is to begin all over again with the logical implication of Descartes' unification of modern physics and physiology. Once more, from Chapter IV above, this is the picture that every introductory text for 'scientific psychology' asks students to take for granted in the current year, 2005. It uses the naïve-realist, 'physical mode of thought,' that yields the conclusion, "Physics, if true, shows that naïve realism is false." We do not sense material bodies, not even our own.

We come into this world belief-less. We — at least our bodies — are bombarded immediately by heterogeneous stimuli — light, sound, odors,

215

I See Colors

heat, etc., coming from the environment where they are jumbled together. First, they are filtered by our five sense organs. Eyes pick out light, ears filter out sound, skin responds to kinetic energy, resistance, and so on. The different or heterogeneous stimuli, after being filtered, trigger non-differentiated or homogeneous electro-chemical chain-reactions conveniently lumped together under "nerve-impulses." Only these homogeneous nerve impulses, traveling via afferent neurons, reach the brain, at which point they correlate with . . .

Now, add Chapter IV's description of the Rutherford view of the physical world, the one with which Eddington's Table #2 is consistent. It is difficult to know for certain what Einstein's view of subatoms would be, since he clung to his 'field' theory to the very end. Regardless, Chapter IV has explained the reasons behind the quintalist commitments on which this book is based.

This is not the place to repeat all of the evidence. A quick summary should suffice. For instance, the cover of this month's *Discover* magazine is headlined with these words: "If an Electron can be in Two Places at Once, Why Can't You? Sir Roger Penrose's answer—page 28." Note the "If." The fact is that no electron or photon can be in two places at once. Only two electrons or photons can. The Brownian effect already proved that about molecules. That is why I trust J. Jeans' claim that, in ordinary air, "each molecule collides with some other molecule about 3000 million times every second and travels an average distance of about 1/160000 inch between successive collisions." (Actually, I trust E. Kasner and J. Newman who reported Jeans' claim in a book whose title could hardly be more appropriate: *Mathematics and the Imagination.*) What's more, only the located-particles theory, not the field-waves theory accounts for the photoelectric effect, Compton's results, and the recent experiments best explained as sending individual photons through an experimental device "one at a time." (*Cf. The New Scientist*, 11 March, 1995, p.18).

Only those in the Bohr camp would think a photon or electron — or a person! — can be in two places at once. Why, in fact, does Penrose believe that he, like other persons, cannot be in two places at once?

216

The Wonderful Myth Called Science

Recent issues of *Discover* magazine have assured us that we each have zillions of clones in this multiverse of ours.

Such absurd claims show how essential is Einstein's wise advice, namely, to understand that "The whole of science is nothing more than a refinement of everyday thinking." If string theorists were right to say there is more than one Penrose in this multiverse, we would have to learn what the other Penroses think about one person's ability to be in two places at once. Suppose we found out that his, I mean their, other answers were different. Which Penrose-answer should we trust? But of course, you and I can only take any of many Penroses' existence on faith.

None of us can begin with any but our own uncloned self. How about you, then? How many of you are there? You can answer "How many of me are there?" only if you understand (i) what the question means and (ii) what the reference of that particular "me" is. (There are two me's in this paragraph, now three; do they all refer to the same existent? I mean, to one existent, not to similar existents.)

The idea of catching up with Socrates and Descartes, which means learning in our turn what they learned ahead of us, is this. Two thousand years before Descartes, Plato collected all of the reasons he could muster in favor of the hypothesis that each of us is a psyche, distinct from the mortal body we presently inhabit. He wove them into his famous dialogue, *Phaedo*. At the end, in order to show the need to adjust our talk to fit our thought, he recounted Crito's question to Socrates: "How do you want to be buried?" Socrates replied: "In any way you like; but you must get hold of me, and take care that I do not run away from you." Then he explained what he meant to the other friends gathered around the place of execution:

> I want you to be surety for me to him now, as at the trial he was surety to the judges for me: but let the promise be of another sort; for he was surety for me to the judges that I would remain, and you must be my surety to him that I shall not remain, but go away and depart; and then he will suffer less at my death, and not be grieved when he sees my body being

burned or buried. I would not have him sorrow at my hard lot, or say at the burial, Thus we lay out Socrates, or, Thus we follow him to the grave or bury him; for false words are not only evil in themselves, but they infect the soul with evil. (Plato, *Phaedo*, trans. Burnet, 115 C-D)

The moral of that speech is clear. To undertake the examined life completely, we must scrutinize the naïve-realist conviction that we are bodies. We must begin with what Descartes singled out as the first thought that a person committed to the examined life can believe is conclusively true: "I exist, and I know I exist right now at least, because I'm right now thinking about it." The next question is, "What kind of being am I?"

That is different from the question, "Who am I?" As the comedian Jackie Mason notes, if we don't remember who we are, we can ask our friends. That's what often sets amnesiacs on the right track, namely, when a family member who sees them on TV recognizes them and goes to 'claim' them. If Socrates and Plato and millions of other people are right about reincarnation (or about just starting all over again, bodiless as all the previous times), even Jackie's answer may not suffice.

This question is "What, not who, am I?" Descartes is very clear about that, as anyone can see who studies the second of his *Meditations*. After noting that he is now absolutely confident that he exists and is actually thinking about it, he goes on:

But I do not yet know clearly enough what I am, I who am certain that I am; and hence I must be careful to see that I do not imprudently take some other object in place of myself, and thus that I do not go astray in respect of this knowledge that I hold to be the most certain and most evident of all that I have formerly learned. That is why I shall now consider anew what I believed myself to be before I embarked upon these last reflections; and of my former opinions I shall withdraw all that might even in a small degree be invalidated by the reasons which I have just brought forward, in order that there maybe nothing at all left beyond what is absolutely certain and indubitable. (R. Descartes, "Principles of Philosophy" *Philosophical Works of Descartes*, Vol. 1. p. 150 trans., E. Haldane and G. Ross)

It is time for would-be scientists, that is, for those who pray to be truly wise, to be more critical than has become the custom among uncritical 'scientists' who have not yet realized the truth behind Einstein's principle, "The whole of science is nothing more than a refinement of everyday thinking." Each must ask "What everyday thinking can I refine?" Each must answer, "Only that which is done by the self I call 'I'."

C. MY ACT OF 'SEEING'

Einstein, Bohr, and scientific observation.

Should we trust that everyone who uses "observe" knows what it means? 'Scientific' psychologists don't agree about it. Fechner, Wundt, and James held that the observation they relied on was introspection. Watson not only denied that such observation existed, he returned to 'pre-scientific' naïve realism and insisted that the only reliable observing is observing organisms and their behavior.

The original linguistic-turning philosophers split into diametrically opposed factions, with some insisting that the only valid observation-sentences (protocol sentences) are those which relate to physical objects, and others arguing that the only valid protocols are those relating to sense-data. (Ayer switched sides.)

The current chaos among physicists pursuing a unification of relativity and quantum theories comes down to observation claims. Einstein wrote as if observing lightning strikes and other distant things were the fulcrum on which relativity rested. Bohr's camp made observing electrons the fulcrum on which they rested their statistical indeterminacy principle. Russell's Einstein-approved claim is that both observation claims are false.

219

I See Colors

Since, like us, everyone gets most of their 'higher learning' from reading, the theory argued for here is based on the premise that each reader must solve the perception-problem personally. A convenient way to do that is to answer the crucial question, "What do we readers observe?" Do we observe books? Texts? Colors? No reader can say with confidence what they observe unless they know what "observe" means. And no reader can say what "observe" means who is not certain what things they observe. The two questions and their answers are inextricably interdependent.

This brings us back to the earlier question. How do we learn what those words (which are not words) mean? To grasp the full dimension of that question, it helps to think about related questions. How can anyone discover that they have suddenly gone blind but not deaf? How do people born blind discover they are blind and not deaf, that they cannot see but can hear? They cannot see their hearing nor hear their inability to see. And what about those born both blind and deaf? Is it really possible for them to know they cannot see or hear?

If it is possible to zoom into one of the most far-reaching questions any of us can ask, then "How do we get the concepts needed to know what those questions mean?" should be at the top of the list. Einstein said, "Not from sense experience." But, to justify our belief that solipsism is false and that other things exist, we need to believe there are concepts that can get us beyond sense experience.

> A few more remarks of a general nature concerning concepts and [also] concerning the insinuation that a concept—for example that of the real—is something metaphysical (and therefore to be rejected). A basic conceptual distinction, which is a necessary prerequisite of scientific and pre-scientific thinking, is the distinction between "sense-impressions" (and the recollection of such) on the one hand and mere ideas on the other.

> There is no such thing as a conceptual definition of this distinction (aside from circular definitions, i.e., of such as make

a hidden use of the object to be defined). Nor can it be maintained that at the base of this distinction there is a type of evidence, such as underlies, for example, the distinction between red and blue. Yet, one needs this distinction in order to be able to overcome solipsism. (A. Einstein, "Reply to Criticisms," p.673. P. Schilpp ed. *Albert Einstein: Philosopher-Scientist*)

Russell's Einstein-approved thesis undermines relativity-theory.

The postulate that the distinction between what we sense and what we can know by conceptual theory is a valid distinction is fundamental to Einstein's relativity theories. In *Relativity: the Special and General Theory*, he invoked this distinction as critical for his far-reaching departure from common sense.

Before you read what he wrote, answer one simple, everyday-thinking question. When you go to the theater to see a movie, can you see the picture and hear the sounds at the same time? (Don't read on until you have tried to answer that question.)

> The concept [of simultaneity] does not exist for the physicist until he has the possibility of discovering whether or not it is fulfilled in an actual case. We thus require a definition of simultaneity such that this definition supplies us with the method by means of which, in the present case, he can decide by experiment whether or not both the lightning strokes occurred simultaneously. At long as this requirement is not satisfied, I allow myself to be deceived as a physicist (and of course the same applies if I am not a physicist), when I imagine that I am able to attach a meaning to the statement of simultaneity. (I would ask the reader not to proceed farther until he is fully convinced on this point.) After thinking . . . (A. Einstein, *Relativity: the Special and General Theory*, Chapter 8, "On the Concept of Time in Physics"]

"Decide by experiment" means observing or seeing. The context for this passage is the famous thought-experiment he used to introduce

I See Colors

Chapter 9, "The Relativity of Simultaneity." A train is traveling from, say, Worcester to Boston. Aboard the train is observer #1. In an automobile near Framingham, observer #2 sits motionless, waiting for the train to pass. A severe thunderstorm is in progress, and a double bolt of lightning strikes the tracks, one bolt behind and one in front of the train. The man in the train happens to be leaning out of a window and, with the help of a mirror, is able to see both lightning bolts strike the rails. The man waiting in the car sees them both as well. Question: "If the lightning bolts are simultaneous according to one observer, will they also be simultaneous according to the other?"

Einstein gives two answers and switches them back and forth in order to protect his desired conclusion. The two answers hinge on a distinction between a coordinate system used to specify distances traveled by light, on the one hand, and the private effects experienced by different 'observers.' Coordinate systems used to specify distances traveled by light involve conceptual guessing about things and events not experienced. A decision about private effects experienced by an 'observer' involves the observer's conceptual thinking about what he or she directly experienced.

In order not to confuse the two, the first and obviously true answer is that no one has ever seen a railroad track, a train, a car, an embankment, or light of any kind whatsoever. Period! Not even if they (rather than swarming subatoms alone) exist. Nor will any person ever see another person. Or what another person sees. Or what another person thinks about what exists, what is seen, what is heard, or what is felt. Period!

The only thing anyone ever sees — or visually experiences[*]— are effects private to him or her. Unless Einstein did not knowingly agree

---

[*] The effects are not literally sensed with — non-existent! — eyes

with Russell's "marvelously concise" and "masterful" formulation of the falsity of naïve realism, that must be the foundational reply to Einstein's question about the meaning of "simultaneity." Everything else must be thought out relative to that principle.

The error of naïve realism

The conclusion that follows from the falsity of naïve realism is crystal clear: *only thinking enables us to learn about anything not subjective to us.*

Thinking, not sensing, is needed to overcome solipsism. Believing in bodies of any kind — or other people! — is an act of faith. In relation to them, each of us is like someone guessing what is inside a watch that we can see only from the outside. Each of us is even better pictured as being inside the watch trying to guess what's outside. Best of all, each of us can be imagined as inside a closed elevator trying to guess what's outside. The only instruments we have for making our guesses are the theories we create inside, the same inside where we find those indispensable clues, the utterly real sense-effects, caused (we think) by never-observed outside bodies.

Einstein's and Bohr's radical inconsistency regards observation.

Conclusions about physical bodies — e.g., about stars, fossils, DNA, etc. — that is, about bodies independent of us and our observation of them, are supposed to be 'decided by experiment,' that is, by observation. What can 'scientists' observe? What can we who read about their conclusions observe?

When we find Einstein implying that lightning bolts can be observed, we are compelled to ask whether Einstein forgot that he agreed with Russell's rejection of naïve realism. Was his agreement all along just an abstract intellectual act, akin to assenting to Euclid's proof that, once we know about opposite interior angles of a line that bisects two parallel lines, we can deduce that every triangle's angles must equal a straight line, that is, 180^0?

I See Colors

There is evidence that such was the case. First is Einstein's statement, "The overcoming of naïve realism has been relatively simple." Nothing is farther from the truth. Consider secondly his follow-up comment on Russell's synopsis of the reasoning which shows that naïve realism is (partly) false. "Apart from their masterful formulation these lines say something which had never previously occurred to me. For, superficially considered, the mode of thought in Berkeley and Hume seems to stand in contrast to the mode of thought in the natural sciences." Yet it was precisely Berkeley's and Hume's deeper analyses of Descartes' causal reasoning that caused (!) Kant to wake from his dogmatic slumber and invent his revolutionary theory that begins with the non-derivation of concepts from experience. Einstein didn't made the connection till 1949?!

But the most convincing piece of evidence is found in a question Einstein asked only five years before his death at the age of 76. The question goes to the very core of this book's central question, "What do readers see?" It forms the topic of the first paragraph of perhaps the most cautious of the Einstein biographies, 'Subtle is the Lord . . .' , written by a physicist, Abraham Pais. The author clearly regarded the question as an extremely significant one. Otherwise why would he use it for the very opening of a 500-page biography of Einstein?!

> It must have been around 1950. I was accompanying Einstein on a walk from The Institute for Advanced Study to his home, when he suddenly stopped, turned to me, and asked me if I really believed that the moon exists only if I look at it. The nature of our conversation was not particularly metaphysical. Rather, we were discussing the quantum theory, in particular what is doable and knowable in the sense of physical observation. The twentieth century physicist does not, of course, claim to have the definitive answer to that question. He does know, however, that the answer given by his nineteenth century ancestors will no longer do. (A. Pais, *'Subtle is the Lord . . .'*, p.5)

Does the moon exist only if I look at it?! No one sees it even when they do look toward it. The moon has never been seen. Why did Einstein ask a question that assumes the naïve realism whose falsity he praised Russell for summarizing?

The answer is clear. When he was wrestling with Bohr and Bohr's colleagues, he was not thinking about the evidence for his own relativity theories. He knew they knew it was never possible to see electrons, only to infer them. He was always looking for a way to prove that facts about electrons would never depend on us or our observing, just as in everyday thinking no one believes the moon depends on us. Yet that is what Bohr's side argued. Recall the passages cited in Chapter IV above.

> ...The questions with which Einstein attacked the quantum theory do have answers; but they are not the answers Einstein expected them to have. We now know that the moon is demonstrably not there when nobody looks. (N. D. Mermin, "Quantum Mysteries for Anyone," in Cushing and McMullin, p.50) "The Universe as Hologram: Does Objective Reality Exist or Is the Universe a Phantasm?", by M. Talbot, in the 9-22-87 *Greenwich Village Voice.* "Quantum Weirdness: Physicists are wondering whether a tree—or anything else—must be observed before it really exists," by M. Gardner, in the October, 1982 *Discover* magazine. "Quantum Theory and Reality: The doctrine that the world is made up of objects whose existence is independent of human consciousness turns out to be in conflict with quantum mechanics and with facts established by experiment," by B. d'Espagnat, in the November 1979 *Scientific American.*

That fundamental premise of the Bohr camp is why Einstein objected to their interpretation of quantum effects or events. "The belief in an external world independent of the perceiving subject is the basis of all natural science," Einstein had written in his 1931 essay entitled "Maxwell's Influence on the Evolution of the Idea of Physical Reality." In 1949, he repeated it. He explained the contrast between the Bohr camp's statistical, probabilistic, undetermined-prior-to-observation, hence altered-by-observation, interpretation and his own view as explicitly as he could:

I See Colors

What I dislike about this kind of argumentation is the basic positivistic attitude, which from my point of view is untenable, and which seems to me to come to the same thing as Berkeley's principle, *esse est percipi*. "Being" is always something which is mentally constructed by us, that is, something which we freely posit (in the logical sense). The justification of such constructs does not lie in their derivation from what is given by the senses. Such a type of derivation (in the sense of logical deducibility) is nowhere to be had, not even in the domain of pre-scientific thinking. The justification of the constructs, which represent "reality" for us, lies alone in their quality of making intelligible what is sensorial given (the vague character of this expression is here forced upon me by my striving for brevity). . . .

Roughly stated the conclusion is this: Within the framework of statistical quantum theory there is no such thing as a complete description of the individual system. (A. Einstein, "Reply to Criticisms," p.669, 671. P. Schilpp ed. *Albert Einstein: Philosopher-Scientist*))

But Einstein's inability to be consistent with Russell's "masterful" refutation of naïve realism clearly poisons his own relativity theories. No one showed this more straightforwardly than M. Gardner whose exposition of Einstein's relativity theories is the best.

We are driven to conclude, therefore, that the question of whether the flashes are simultaneous cannot be answered in any absolute way. The answer depends on the choice of a frame of reference. Of course, if two events occur simultaneously at the same spot, it can be said absolutely that they are simultaneous. When two airplanes collide in midair, there is no frame of reference from which the smashing of both planes will not be simultaneous. It is important to understand that this is not just a question of being unable to learn the truth of the matter. There is no actual truth of the matter. There is no absolute time throughout the universe by which absolute

The Wonderful Myth Called Science

> simultaneity can be measured. Absolute simultaneity of distant
> events is a meaningless concept. (M. Gardner, *Relativity for the*
> *Million,* p.43; & The Relativity Explosion, p.45)

That is simply wrong. There is no time, not anywhere. There are no frames of reference apart from thinkers who can pretend them the same way we pretend one-dimensional lines, two-dimensional planes, etc. There are literally no events and no airplanes. Even if there were, why think that, when two planes are involved in a single crash, it makes two events occurring simultaneously? His concluding "Absolute simultaneity of distant events is a meaningless concept" is the exact type of positivism Einstein rejected in his case against the Copenhageners.

The clinching clue that shows Einstein's relativity to be trapped in objective indeterminacy is that courageous admission by Gardner that "There is no actual truth of the matter." Given the finite velocity of light that all of us today posit as essential for seeing anything, it is impossible for any observer in any imaginary frame of reference (referentiality, again) to see plane collisions or lightning strikes, which makes it impossible to see external events occurring simultaneously (non-successively) or otherwise. The occurrence of effects allows inferences about external causes, but the causes are independent of the effects, and the concept of simultaneity is created, not derived from the effects. "The observer, when he seems to himself to be observing a stone, is really, if physics is to be believed, observing the effects of the stone upon himself." According to Gardner and (here) Einstein, there is no truth of the matter answer to the question, "Do independent-of-us stones or moons or planes even exist?"

Self-contradictions or inconsistencies in extraordinarily complex belief-systems are easy to miss if one is already a long-standing believer. A very long time ago, I tucked a now-yellowed book review of Loren Eisely's *The Mind as Nature* into his book, *Darwin's Century*. The review begins:

> There is a legend circulating about a late distinguished
> scientist who, in his declining years, persisted in wearing

enormous padded boots. He had developed a wholly irrational fear of falling through the interstices of that largely empty molecular space which common men in their folly speak of as the world. ("The Importance of Reverie." The review is unsigned.)

Boots won't help. They are as porous as feet, ankles, torso, etc., all those things which common-sense folk imagine as their body. As for today's relativity and quantum experts, they are no different from Einstein who never saw lightning or the moon, from Bohr who never saw an electron or an electron microscope, or from the two of them who never even saw each other.

Hume's experiment.

As we inch toward a final answer to, "What do readers observe?", it helps to reflect that Einstein's inconsistency shows how hard it is to overcome our deeply-entrenched naïve-realist thought-habits. The difficulty is not with the theory. Even Hume thought the theory was clear. Once again:

> But this universal and primary opinion of all men is soon destroyed by the slightest philosophy, which teaches us, that nothing can ever be present to the mind but an image or perception, and that the senses are only the inlets, through which these images are conveyed, without being able to produce any immediate intercourse between the mind and the object. The table, which we see, seems to diminish, as we remove farther from it; but the real table, which exists independent of us, suffers no alteration: it was, therefore, nothing but its image, which was present to the mind. These are the obvious dictates of reason; and no man, who reflects, ever doubted, that the existences, which we consider, when we say, this house and that tree, are nothing but perceptions in the mind, and fleeting copies or representations of other existences, which remain uniform and independent.

The Wonderful Myth Called Science

Hume was mistaken, of course, in his cavalier claim that "no man who reflects" will ever doubt what he, Hume, didn't doubt. Most reflectors in 2005 do whatever they can to do just the opposite of Hume and Kant. They agree with Heidegger that the scandal of philosophy was not what Kant said, namely, that no one has yet proven the external world. The real scandal, they say, is that anyone ever thought it was necessary. That makes it imperative for every twenty–first century learner to decide which is the true scandal. Or whether there is no scandal, just impossibility.

Before he threw in the towel and admitted that he couldn't figure things out, David Hume made certain that every one of his readers would feel the full impact of the modern discoveries. Contrary to those who say that 'philosophers' do not rely on mathematics or experiment the way 'scientists' do, Hume the philosopher-scientist proposed a small test for all of this. It comes in the midst of the paragraph that precedes the passage quoted above.

> I need not insist upon the trite topics, employed by the sceptics in all ages, against the evidence of sense; such as those which are derived from the imperfection and fallaciousness of our organs, on numberless occasions; the crooked appearance of an oar in water; the various aspects of objects, according to their different distances; the double images which arise from the pressing one eye; with many other appearances of a like nature . . . There are other more profound arguments against the senses, which admit not of so easy a solution. (D. Hume, *Enquiry Concerning Human Understanding*, Sec. XII, Pt. I)

". . . the double images which arise from the pressing one eye." This is a far more valuable exercise than Hume realized. It suggests a three-step test which will clarify what is altered and what, being independent of us, 'suffers no alteration.'

Step one must be done in a situation in which the experimenter seems to experience a companion able to participate. If on some cloudless evening the two observers put their books down, go outside the library to

229

look at the moon high in the dark sky, they can compare what happens when only one presses an eye quickly and repeatedly. The other person can watch to see whether or not anything happens to the moon. (Readers who have never done so are invited to see what happens when one eye is pressed quickly and repeatedly.) When they are finished taking turns and dialoguing about what they experienced and which of the dozens of sense-perception theories they believe will explain what they experienced, they are ready for the next two steps of the experiment.

For step two, they return to the well-lit library. They put a book on the table and repeat the experiment done vis-à-vis the moon. The book will not jump (suffer alteration), but what the book-users see — unless they suddenly go blind — will.

For the third and final step, they each look at one of the other one's eyes and take turns pressing one of their own eyes quickly and repeatedly. If the moon didn't jump (suffer alteration), but only what the eye-pressers saw, then other persons' eyes don't jump, only what the eye-pressers see.

The only conclusion consistent with that tri-step experience is the one firmly adopted here. Moons, books, others' eyes, and other people are not the altering, jumping, changing things we see. Hume's wording is good. We see images, representations, and perceptions, things he officially named "impressions," i.e., our sense-data. Kant's wording, "We see the moon, books, others' eyes, just not 'as they really are'," is confusing. We do not see those things, period! They do not even exist, only subatoms. What we see are real sense-data, utterly distinct from moons, books, or eyes. Russell was right. Einstein was right that Russell was right. Berkeley and Hume were right even earlier.

Two additional notes. First, the same analysis must be carried through for each of the other types of sense-objects. When we jiggle an eye, we do not feel the finger or unseen and nonexistent hand or eye-movements. Phantom-limb fingers, virtual-reality eyes, parts of an entire phantom-body are what we feel!

The Wonderful Myth Called Science

Second, that experiment relies — as Hume, the empiricist, said — on "the obvious dictates of reason." By thus destroying "the universal and primary opinion of all men" still trapped in their youthful naïve realism, it helps us put the group-abstractions of anthropologists, historians, sociologists, social psychologists, political scientists, and other 'group' thinkers, into a true-science perspective. It confirms the fact that group-concepts are mental creations, 'shorthand' for simultaneously thinking about many never-seen, hence inferred, individuals. It eliminates the need for such obfuscating ambiguity as the following.

> The principal thesis of the sociology of knowledge is that there are modes of thought which cannot be adequately understood as long as their social origins are obscured. It is indeed true that only the individual is capable of thinking. There is no such metaphysical entity as a group mind which thinks over and above the heads of individuals, or whose ideas the individual merely reproduces.... Strictly speaking, it is incorrect to say that the single individual thinks. Rather it is more correct to insist that he participates in thinking further what other men have thought before him. (K.Mannheim, *Ideology and Utopia*, tr. L.Wirth & E.Shils, ch.I)

Is applying Ockham's Razor by recognizing society, groups, and culture as fictions a good example of 'thinking further'? Yes. Therefore we apply it to Thomas Kuhn's revolutionary work, *The Structure of Scientific Revolutions*, a classic work whose central paradigm is the group construct called "the scientific community."

> In this essay, "normal science" means research firmly based upon one or more past scientific achievements, achievements that some particular scientific community acknowledges for a time as supplying the foundation for its further practice. (T. Kuhn, *The Structure of Scientific Revolutions*, p.10)

"*Its* further practice"?! Kuhn's personal theory — a theory that Einstein would say was created by Kuhn's own individual mind —

231

I See Colors

started a 'scientific revolution' against the myth that 'science' is the name for an evolving 'thing,' namely, a collective body of true, impersonal knowledge constantly being added to and purged of error. The history of individual reader-responses to Kuhn's thesis — viz., that the history of 'science' reveals revolutions rather than evolution — shows that neither of those useful fictions, the evolution or the revolution paradigm, is entirely adequate for capturing the full truth about the intellectual growths and developments of many individual learners over the course of time. A combination of the two paradigm-abstractions was used in a speech given on November 25, 1985, at Moscow State University, by G. Piel, Chairman, *Scientific American.*

> . . . in a world transformed by the application of scientific knowledge, people put that knowledge in the same category with what they know by revelation or other received authority. What needs to be understood is how, scientifically, we come to know.
>
> Scientists know nothing for certain. The advancement of science is a social process, a public process, and yet an intensely private one. Societies that would enjoy its material benefits must understand science in both its aspects. (G. Piel, "The Social Process of Science," in *Science*, Jan. 17, 1985, p. 201)

That editorial excerpt of Piel's speech came to my attention through what die-hard group-thinkers might insist was a social process. But it was an individual colleague who passed an individual Xeroxed page to this individual author.

The Wonderful Myth Called Science

Uncritical thinkers often protest that they can see their own selves, faces, or eyes in a mirror. Hume's test can be used for that as well. Watch your 'self' in the mirror while you press your eye the same way as before. What you will see will jump. You should be able to tell whether you did as well.[*] If you press your eye slowly the way Hume suggested, you will see two of your 'self.' It is easy to become adept at simply crossing one's eyes — no hands! — and doubling everything except the absolute-space size of one's total visual field.

Mathematics is essential in all of these reasonings. It is possible to take a large photo of one's self to the mirror and to compare it with one's self being seen in the mirror. The part in one's hair, one's left shoulder, etc., are on different sides of one's two selves. There is also a self which sees both, and which, together with the two seen selves, yields a total of three selves. Were it not for Ockham's Razor, one's one self would lapse into an MPD. (Multiple Personality Disorder) Or a DID?

The library has even more books and journal articles in 2005 than it had in those 1960's dissertation-writing days. Whoever wishes to learn more of the endless debates about endless details on every topic related to the question, "What do readers see?" will find enough material to spend a century or two reading it all.

Suppose that the experiments suggested above convince you that you do not get your concept of your eyes from seeing them, nor your concept of seeing from seeing seeing. A fortiori, you did not acquire your concept of your brain by seeing, hearing, smelling, tasting, or touching it. Nor have you ever seen another person's brain. How did you acquire your concept of your brain? What made you create a concept of it as a part of

---

[*] Do ethnologists really imagine that mirror gazing chimps and elephants can think such complex 'self-knowledge' thoughts?

you or your body, distinct from those other parts, viz., your eyes, your ears, your nose, your tongue, and your skin? How do you square your concept of your brain with Rutherford's theory about subatoms?

"Think movie!" Once again, switching mindsets.

If you ever took a course in Shakespeare's tragedies, you may have read the play, *Romeo and Juliet*. If you did that, or if you saw the movie version, then try to remember which of the two died first. Was it Juliet who stabbed herself to death or Romeo who drank poison?

As noted earlier, every normal person initially finds arguments against naïve realism deeply flawed, and patently self-contradictory. For instance....

How can any reasonable person say eyes do not exist, then describe in detail the experiments which make no sense if eyes do not exist? How can anyone talk about ears, nose, tongue, and skin, if they do not exist? What sense does it make to invoke Jupiter, Jupiter's moons, Olaus Roemer's telescope, the distance from Jupiter to earth, volleys of electro-chemical discharges via afferent neurons from eyes to brain, if (i) none of those things exist, if (ii) their imaginary interactions are used as part of the proof that they have never been observed, and if (iii) that negation becomes an integral component of a unified theory about reality?

The reply begins, "Think movie!" Self-consciously notice what we do when we go from 'the real world' into the 'imaginary world' of the movie, and what we do when we return to 'the real world.' Then generalize what happens and use it to explain literally dozens of our life experiences.

What do we do? We switch mindsets. Movie-going illustrates a profound type of mindset-shifting. Consider the movie version of *Romeo and Juliet*, for instance. Common-sensically, it seems obvious that the only sensations that are important to movie-goers are the two-dimensional 'moving' picture on the screen and the sounds from the loudspeakers. If we switch into a 'science' mindset, we explain the movie-going experience with ideas about physics, photo and audio technology, and biology (eyes, ears, afferent nerves, and brain). If we

The Wonderful Myth Called Science

switch to a 'liberal-arts' mindset, we ignore the 'physics' and discuss the plot, the characters, the similarity to *West Side Story*, the genius of Shakespeare, and enough other things to fill a thousand library books. But those are pictures of Romeo, not Romeo on the screen, and those are sounds, not Romeo's voice from the loudspeaker.

During our naïve days, we conclude that the 'science' story is about reality, the 'liberal arts' discussion is about fiction. (The answer to the opening question is that neither died. Neither ever existed.) There is the physical reality that we pay no attention to during the movie. The 'world' of the Montagues and Capulets is created, unnoticed by our imagination. This is what Einstein emphasized. Our mind supplies most of 'experience.' Take away all memory, reduce the reader or movie-goer to the mental state of a newborn, and the only thing left of that individual's stream of experience will be a meaningless sequence of sights and sounds and other sensed 'effects' inside us that exist for only an instant and are succeeded by new ones. But movie-goers are not newborns. They have a vast memory base, relevant gobs of which are instantly activated in unbroken succession by the movie's sights and sounds.

Now, however, we must push on to a deeper truth. While we are watching the movie, we can 'step back' at any given moment and think "I'm not a body in a cushioned seat in a theater using eyes to see pictures on a screen produced by an unseen projector or ears to hear sounds. . . No. At most there are zillions of individual subatoms 'obeying' dozens of laws of physics, chemistry, biology, neurology, etc. Russell and Einstein were right, I am experiencing effects in me, and I am now switching to a third mindset, that of my ultimate unifying theory."

We constantly switch mindsets. Now we must notice it, get used to it, and then construct a unifying theory, one we can use to account for all the others and it itself. Equipped with a self-conscious habit of thinking about switching mindsets, we can apply it to our everyday-thinking lives, from reaching to shut off the alarm in the morning to closing our eyes again at night. Those lives are comparable to being at a multi-channel movie, complete with color, sound, odor, taste, etc. What is really going on is not at all what we naïve-realistically grow up believing is going on.

235

I See Colors

In the words of Jonathan Harrison, our daily experience is comparable to a constant hallucination. So long as it is consistent, 'it seems as good as reality.'

Agree or disagree, no one can understand Einstein who does not grasp the psychology of switching from a naïve-realist mindset to the virtual-reality, consistent-hallucination mindset, and back. Doing so is easier for us than for any of our ancestors. Movie-goers and TV viewers are used to thinking about 'virtual reality.' After all, anyone who can 'follow' the Star Trek idea of holodecks can understand it. A full-bodied, solid hallucination, consisting of sense-data, is the same thing.[*]

Summing up! When we are in the midst of a movie, we can instantly switch from one way of thinking to the other and back again. We must practice doing the same thing while in the midst of our everyday-thinking experience.

What about swarming subatoms, if they exist?

What is really going on out in the world of subatoms and photons, even lightning-strike photons, if they exist?

First, understand that the physical 'world' has no beauty whatever, if by "beauty" we mean something comparable to what we sense. It is pitch dark, utterly silent, and devoid of all sensible features such as taste,

---

[*] Note. As should be expected, whoever thinks there is agreement on what constitutes hallucination can begin his or her disillusionment — or enlightenment — by more reading. *Origin and Mechanism of Hallucinations:* Proceedings of the 14th Annual Meeting of the Eastern Psychiatric Research Association, held in New York City, Nov. 14-15, 1969, edited by W. Keup, is a good start. See also "Extreme States," by S. Kotler in *Discover,* the July 2005 issue.

smell, heat, cold, pain, tickles, and pleasure. Whatever beauty nature-in-itself possesses is accessible only to our intellect.

Then imagine doing what Gardner would say is impossible. Stop the world the way we imagine stopping a merry-go-round to get off. The physical universe is a vast void occupied solely by subatoms which do only two things: rest or move. The idea behind Descartes' coordinate system is that, if we could stop the world of bodies dead in its tracks, we could map out the precise location of every subatom vis-à-vis every other. If we add Newton's view of light as photons, tiny billiard balls, they too could be identified by the three coordinates symbolizing their precise location vis-à-vis every other subatom.

Total relativity means that we can arbitrarily — not irrationally! — select any subatom as the anchor for our purely imaginary coordinate system. However, so long as every body remains frozen in place, no clocks will be running. Is it necessary to point out that clocks are not time, or that a slower-running clock is not slower-running time as well?! There will be no motions of any kind to describe as uniform, accelerated, decelerated, etc. Just as there is no field of force to guide the cannon ball (if Sutton is right, "there are no 'wave mechanics' in external ballistics"), there are no fields of force — and above all no waves of probability! — to guide billiard-ball photons as soon as we press the button and those unseen subatoms begin once again to change positions vis-à-vis each other. Only then will we need the purely imaginary 'fourth' dimension we call "time" in order to describe where things were (but are no longer), are (for an instant), and will be (but are not yet).

How could this be used to explain the Michelson-Morley experiments? First, it makes absolutely no sense to say something is moving unless we ask "relative to what?" Think critically about popularizers' explanation-attempts. Second, check the evidence that might show that the guess of Roemer — "Photons move 1000's of miles per second from the point where their source of radiation or reflection vis-à-vis every other body was at the instant they began traveling to earth" — was true.

I See Colors

Besides subatoms, are space or time real? The answer is "No!" And adding two fictions together does not yield one reality. Every library should have anthologies that gather together the endless theories about space, time, and space-time created by imaginative thinkers from Plato on. Many will have essays about the paradoxes created by the idea that space and time have now fused into a four-dimensional space-time continuum. All of them should but won't contain the following report:

> Paul Langevin was one of the first who protested against calling time "the fourth dimension of space." Einstein himself admitted that the asymmetry of time is preserved even in its relativist fusion with space when he recognized that "we cannot send wire-messages into the past." When Meyerson in the session of the French Philosophical Society of April 6, 1922 insisted on the distinction of space and time even in the theory of relativity, Einstein, who attended the session, explicitly agreed. Meyerson's argument was fully developed in his book, La Déduction Relativiste, and Einstein in his highly favorable comment about it again praised Meyerson's criticism of the spatialization of time. According to Einstein, the spatialization of time is a misinterpretation of the theory of relativity, a misinterpretation committed not only by popularizers, but even by many scientists, though it is often present in their minds only implicitly. (M. Capek, "Relativity and the Status of Space," p. 172.)

Einstein created his field theories, not to answer "What exists?", but to answer "Why do things behave as they do?" But that is question #3, and it makes no sense to ask it until questions #1 and #2 have been answered. So we begin with, "Do physical bodies, billiard balls, subatoms, fields of probability waves, etc., exist?" Only if some part or parts of that question are answered "Yes," does it make sense to ask question #2, "What do those existents do?" If anyone believes moons exist even when no one is looking, they must answer "What do moons move in relation to?" before guessing about "Why do they move?" M. Kline summed up the chaos of answers to that question, answers that

The Wonderful Myth Called Science

range from the force of gravity to gravity waves, gravitons, and gravitinoes, or to warps in the fabric of space-time, with his comment that "The greatest science fiction stories are in the science of physics." (M. Kline, *Mathematics and the Search for Knowledge*, p.122)

One thing is certain. Moons cannot literally move in relation to 'systems of reference' or 'coordinate systems.' Those creations of our imaginations have no matching counterparts in reality.

However, even if the 'world' were stopped, frozen in its tracks, our stream of conscious experience would go on. Its three tributaries are not caused by any physical bodies. And we would still need our complete thoughts about imaginary past-present-future time-lines to describe our ongoing, three-component, virtual-reality, conscious experience.

In fact, it is entirely conceivable that nothing physical at all exists.

We need the concept of time, even though time does not exist.

While I was still in the fifth grade, my classmates and I were shown a 16mm documentary movie on the heart. It terrified me. It brought home how totally I depended on that organ continuing to beat 72 times every minute of every hour of every day of every month. There are times now when my mild tinnitus is enough for me to 'take my pulse' by changes in the sound. At night, with my ear on the sheet-covered mattress, I can hear the faint sound of each beat. Then I fall asleep. Did Einstein ever wonder, "Does my heart exist only if I hear it?"

For anyone still worrying about Einstein's claim that we cannot have an idea of simultaneity if we can't 'fulfill' it, two things help. First, understand that he was resorting to the same positivism he rejected when he thought about Bohr's theory. He needed it to smooth over the 'no truth of the matter' inconsistency in his field theory. Second, compare seeing to hearing. Both are observing. For instance . . .

Are you able to read this book and listen to Beethoven's Ninth Symphony simultaneously? Can you see what you see right here — HERE! — and hear sounds at the same time that you are also pondering the question, "Am I sure that the seeing I am doing is really not the same

I See Colors

thing as the hearing that I am doing, that is, am I positive that being blind but not deaf is really different from being deaf but not blind?" Can you then choose the true answer to that question without stopping your seeing and hearing? While seeing and hearing, can you also ponder this question: "What proof do you have that you did or did not exist fifteen minutes ago?"

How many times in the previous paragraph did you have to use the idea of simultaneity? Regardless of what the 'text itself' said to you, did you read "at the same time" as the writer's way of evoking the same idea? Ask the same question re "without stopping" and "while . . . also." No one who does not already have common-sense ideas about time, a long time, a brief time, no time at all, and the same time can even read Einstein's paragraph. Physicists are free to use old words with opposite meanings, as is the case with "straight," "instant acceleration," "curved space," etc. The result is that, when they use such terms for their inferences (guesses) about never-observed physical bodies, they confuse not only the naïve, but themselves as well.

Be cautious about experts' observations about what is and what is not observed. According to some, we can gaze at stars in the depths of space and simultaneously! be gazing at time, since we see the stars, not as they are right now, but as they were in the past. Not so. Ian Shelton did not see into the past when light from supernova 1987A reached earth on February 23, 1987. He observed an effect in him produced that night by light that began a trip to earth 160, 000 years earlier. Yet experts who ought to know better defend claims that the universe is very old by the absurd — and confused — idea that 'we can see stars that are billions of light years away, which proves the light had to have been traveling for billions of years.' And don't be surprised if writers, challenged to defend that absurdity, offer the defense that there is no single 'now' everywhere in the universe.

True, there isn't. But that is because, in the same way there is no Santa Claus anywhere in the universe, there are no 'nows' anywhere, either. Nor any past. (Where would it be?) Nor any future. (Where is it?) All the room in the universe is filled with what exists now, in the present. But there is no present, either, only the things that exist presently or now. Apply Ockham's Razor. Things that are now were not always as they are now, and they will not always remain as they are now, but the only way they are is the way they are now. And now. And . . .

Remembering, imagining, and expecting are mysteries. How do we acquire ideas of again, past, ages ago, the time before any humans existed, yesterday, today, right this instant, the present, right now, now, here and now, later, tomorrow, next year, etc.? (Did you notice the switch from "Stop here" to "stop now" at the beginning and end of Segment A of Chapter III-1) If a year is 365 days and "a day" = "the time it takes earth to spin once," what was a year before the earth came into existence about four and a half billion years ago? Etc.

Time does not exist. But we would be lost without the concept(s). Our whole science of the physical world rests on the intelligibility of our 'time' concept(s). And we must use a picture of time (spatially) 'spread out' in durations that we mentally divide into segments, large (eons, light years) and small (nanoseconds). But this 'spread out' does not mean simultaneously co-existing. How many of your heartbeats can you hear simultaneously? How many pages can you read simultaneously? How many bars of Beethoven's Ninth can you enjoy simultaneously?

If I wanted to find Beethoven's Ninth, where in the universe would I look for it? Ask yourself which of the following answers is the true one. I can find it in the collected pages of 'a printed score.' I can't find it there, because those are silent ink marks. I can find it on eight 78 rpm records. I can't find it there, because those are just disks with uneven grooves. I can find it on a cassette tape-recording. I can't find it there, because that is just a long ribbon of plastic film coated with particles of magnetized iron oxides. I can find it in a symphony hall where it is actually being played. I can't, because I can only hear successive effects inside me, though each actual effect is experienced simultaneously with present memories 'of' past heard-effects and, if I have 'heard' the Ninth often enough, presently-awakened expectancies 'of' still-unheard-effects. There literally is no such 'thing' as Beethoven's Ninth, though there are those effects, memories, anticipations, pleasures, and the thoughts I have just been understanding 'about' it. *

Science and the laws of nature

Question: What is a law of nature?

Answer: A very useful fiction. And yet, for many thinkers, "science" is practically a synonym for knowledge of the laws of nature.

It takes very little reflection to recognize the purely metaphorical nature of 'a law of nature.' First, we latecomers get our idea of laws the same way we get our idea of Santa and science. Under the impetus of hearing "law" over and over and then hearing 'talk' that uses the 'word,' we create our concepts of 'laws.' We learn about traffic laws, tax laws, libel laws, etc., which we are free to either obey or disobey. But bodies, large or small, do not know anything. Hence they know of no 'laws of nature.' Hence, they cannot obey or disobey them. Poetry galore!

Such metaphors or fictions are not only useful they are essential for mental economy. But, unless they are recognized as metaphors, they can seriously impede clear thinking. Darwin, aware that this was the case with gravity, realized that there are similar problems with 'natural laws' relating to natural selection and nature.

---

* An addendum. Even though there is no such real thing as the Shakespeare play, *Romeo and Juliet*, do professors of English literature ever say to students when they come into class, "Today we will be talking about the fictitious Romeo, the imaginary Juliet, and the merely pretended animosity between the pretended Montagues and the pretended Capulets"? Do we adults have to keep reminding ourselves "When I talk about Santa Claus, I must remember that he does not exist?"

The Wonderful Myth Called Science

> It has been said that I speak of natural selection as an active power or Deity; but who objects to an author speaking of the attraction of gravity as ruling the movements of the planets? Every one knows what is meant and is implied by such metaphorical expressions; and they are almost necessary for brevity. So again it is difficult to avoid personifying the word Nature; but I mean by Nature, only the aggregate action and product of many natural laws, and by laws the sequence of events as ascertained by us. With a little familiarity such superficial objections will be forgotten. (C. Darwin, *The Origin of Species,* Chapter IV.)

The best introduction to the correct way to think about laws of nature are the writings of Hume. (His most important contribution will be discussed later.) And the easiest example with which to begin is the well-known law of inertia, "All bodies remain at rest or in uniform motion in a straight line unless disturbed by an outside force." That 'law' does nothing but tell us what bodies do. Every normal human being learns that, if a coconut or a stone is lying in an open clearing, it can be moved by a kick. It moves while our foot is pushing it. But it also continues moving on its own for a short distance. "Why?" is question #3. Also, why does the stone we pick up and throw continue moving after it leaves our hand? Why does the sun move across the sky? Once we have experienced such 'doings' of supposedly-existing things often enough, we expect the same 'doings' the next time we find ourselves in a similar situation. We say "It's what usually happens." That becomes "It's the rule rather than the exception." That in turn becomes "It's a law of nature." But the only realities are coconuts, stones, etc., and our thoughts about what they do.

Here then is a stipulative definition. What we call "laws of nature" do not exist as such. We have generalized thoughts to describe what we can expect things to do in the future, based on what we remember them doing in the past, and we reify that by inventing the metaphor of 'it's a law.' But there are no 'laws' out there in nature.

I See Colors

The bomb which woke Kant from his dogmatic slumber was Hume's careful dissection of the illusion that we actually see or observe anything except before-and-after doings. His analysis exposed the fact that we do not 'inductively' gain from experience the concepts of force, power, energy, or causal impetus we project as the cause of things happening. Even if we assume that coconuts, people, and photons exist, we never see the force of gravity make apples fall, the power of authority causing people to obey, or energy pulling magnets together or pushing them apart. Hume's own best account of the story, told many times in the library's holdings, is found in his *An Enquiry Concerning Human Understanding.*

The all-important distinction between question #2, "What do things do?", and question #3, "Why?" helps to understand the way Newton differs radically from Descartes and Einstein. Like so many before him, Descartes experimented to see whether light diffuses instantly through space as Aristotle said or whether, like sound, it takes time to travel. When he found no evidence that, like sound, light takes time to travel, it bolstered his confidence that he knew why light and other bodies move the way they do. They are being shoved, if not by a large thing (my foot kicking a coconut), then by an invisible sort of fluid. Empty space is not empty.

Newton, at first enthusiastic about Descartes, eventually rejected his idea and pictured stars, planets, moons, and cannon balls as bodies that seem to be moving in empty space, that is, in space not occupied by anything. That is the critical fact about Newton. It is where Newton differs, not just from pretty much everyone before him, but especially from Einstein. The fact that he, Newton, had no clue to how bodies could affect the motions of other bodies across empty space — the Latin was *actio in distans* — did not deter him from spelling out the 'laws' that bodies in outer space appeared to 'obey.' Bodies are what exist physically. Resting or moving is what they do. Why they do what they do, Newton said, "I won't speculate." Newton left question #3, "Why?" unanswered.

The Wonderful Myth Called Science

Einstein, though, was unwilling to say "I don't know either." He rejected empty space and postulated his space-time 'field' theory, a substitute for the old ether. Once we cut through the excess terms and the confusing explanations, what we find is that Einstein returned to Descartes' view. The universe is full, if not full of ether, at least full of a space-time field of force(s). Or a field that substitutes for force(s), e.g., for the forces of gravity and electro-magnetism.

The view in this book is that of Newton. There are subatoms. The moving ones have no contact with each other. There is no empty space between them, because there is no space. To avoid language traps when trying to be crystal clear, we do not say "There is nothing between subatoms," because that might make nothing into something. (Which Bohr's successors now insist is true. They even write about "the difficulty of weighing nothing"!) Instead, we say "They are not touching and there is not anything between them." Still, the thought's the thing, not the (non-) words.

Once the distinction between question #2 and question #3 is clear, it is easy to understand the enormous difference between the law of gravity and the force of gravity named by the law itself: "All bodies attract all other bodies with a force that varies according to distance and mass." Like the law of inertia, the law of gravity tells about the rest and motion of bodies. The law of gravity is part of the answer to "What do bodies do?" The force part is at most a theoretical fiction. It seems to answer "Why at times do they change their doings?" But it doesn't!

> You sometimes speak of gravity as essential and inherent to matter. Pray do not ascribe that notion to me, for the cause of gravity is what I do not pretend to know and therefore would take more time to consider it. (Letter of Newton to Bentley, 1692/93; cited in *Newton's Philosophy of Nature: Selections from his Writings,* ed. by H. S. Thayer, NY, Hafner, 1953, p.53.)

Later commentators were clearer:

I See Colors

The law of gravitation is a brief description of how every particle of matter in the universe is altering its motion with reference to every other particle. It does not tell us why particles thus move; it does not tell us why the earth describes a certain curve round the sun. It simply resumes, in a few brief words, the relationships observed between a vast range of phenomena. It economizes thought by stating in conceptual shorthand that routine of our perceptions which forms for us the universe of gravitating matter. (Karl Pearson, *The Grammar of Science,* London: Dent, 1892, p.99.)

Another complicating challenge to our everyday thinking is this. Even if the gravitational force or warps in a space-time continuum did exist, they would not be answers to question #2, "What do things do?", but to question #3, "What real cause makes apples fall and hot air balloons rise?" Though discovered by asking question #3, they would subsequently have to be added to the other answers to question #1, "What exists?" In the same way, the planet Neptune, discovered by asking "Why is the orbit of Uranus irregular?" is a thing that exists.

In short, nature as a whole is a huge jigsaw puzzle that has required the ingenuity of many great minds to isolate the pieces and then to try fitting them together into *a unified theory covering everything*.

Think movie!

The biggest surprise comes when our "Why?" question about sensation leads to the discovery that, all along, what we were watching were not real coconuts, stones, planets, or other bodies independent of us and behaving in complex ways. These are virtual-reality holodecks you and I are living in, complete with large, screen-size-movie colors, wrap-around sound, smells, tastes, warmth, coolness, hardness, softness, tingling pleasure, tormenting pain, and so on. All of these colors and other real sensed phenomena are used to theoretically create subjective 'worlds' which we hope 'represent' the never-observed objective world.

We have to start re-guessing about the world that is on the outside of the closed elevator we cannot climb out of or inside the watch we can

246

never open. Our 'movie' is our clue to what things are out there, what they do, and why. But what we must eventually try to understand is this mind of ours and how it works. We must also recognize two kinds of laws, physical and psychological.

Physical models: useful but false, pragmatic-fiction pretendings.

First, physical laws. The law of inertia, "All large groups of subatoms remain at rest or moving at a uniform velocity in a straight line unless made to change by something else," is shorthand for so much that we now know if we have been fortunate to have had well-informed teachers or to have read enough of the right books. If we wish to use similar shorthand for the other complex behaviors of subatoms, we refer to the laws of chemistry, the laws of biology, and the laws of neurophysiology.

We must use our naïve-realist concepts as guides when we pretend (falsely) that subatoms form groups. Those original concepts are essential when we wish to pick just certain subatoms out of the whole universe. For instance, you must use your ideas of pages and ink marks when you pretend that, correlated with this virtual-reality 'page' you are right now 'reading,' there are just certain zillions of unseen carbon and other-type atoms making up molecules making up fibers making up 'paper,' and a certain zillion other atoms making up molecules making up 'ink' spread in certain patterns on the 'paper,' etc. Similarly, even though there is no such thing as biological life, we can use the naïve-realist concepts expressed by "the springtime return of foliage to Elm Park" to refer collectively to the dances of untold zillions of just certain of the universe's subatoms.

Shorthand for referring to just certain subatoms parallels the use of mentally-grouped concepts for selectively referring to just certain of the 6.4 billion persons 'making up' the present human population: male vs. female, infant vs. elder, tall vs. short, Republican vs. Democrat, Irish vs. Scot, professor vs. student, diabetic vs. alcoholic, etc. In the same way that the one Einstein belonged to the human race, the male sex, a small-family, the German nation, the physics community, etc., we pretend that a single electron can be part of a hydrogen atom which is part of a water molecule which is part of a glial cell which is part of a brain which is part of a human body which is part of a biological species which is part of the earth's mass which is part of the solar system which is part of the Milky Way Galaxy which is part of the cosmos, etc. Of course, none of those 'physical' concepts represent anything real except the zillions of individual subatoms.

Thus, Einstein's idea that concepts are free creations of our imagination, used in our guesses about the unseen world of physical bodies outside the closed elevator or inside the closed watch, fits perfectly here. Whether to call them logical fictions, pragmatic concepts, theoretical constructs, etc., is a matter of vocabulary. What is important is not the name but the thought. Incidentally, the clarity of Einstein's thinking vis-à-vis the use of freely-created, pragmatically-useful models in physics is nowhere as clearly described as it is in the first two chapters plus the last "Physics and Reality" section of his and Infeld's *The Evolution of Physics*.

Psychological models: useful but false, pragmatic-fiction pretendings

Second, psychological laws. Sense-data, images, and thoughts are as real as physical subatoms. Accustomed as we are to physical bodies that do not depend on us, sense-data strike us at first (and later!) as very strange kinds of things. The regularities 'governing' them are as enormously complex as any physical laws. Noticing these non-physical realities and discovering the regular patterns in their occurrence requires the same kind of imaginatively created models as are needed to discover what physical things exist and how they behave.

That is, Einstein's idea of creative imagination (models) and Kuhn's theory of paradigms (models) must be also be extended to the radically divergent theories of psychologists. If one of the great oversights in the search for a unifying theory is the failure to take seriously Einstein's claim that "The whole of science is nothing more than a refinement of everyday thinking," another great oversight is the widespread failure to realize that the whole of psychological science is also essentially a

The Wonderful Myth Called Science

refinement of everyday thinking. Einstein did not take as much interest in 'the evolution of psychology' as he did in 'the evolution of physics.' However, the central theses of his analysis of 'science' and 'scientific' theorizing are essential to any sound 'science' of psychology. The model-accompanied theory presented in this book illustrates *that* claim and serves as evidence to verify it.

For instance, Einstein's distinction between sense-experience and conceptual theorizing and his theory about the relation of everyday thinking to specialized theorizing are parts of an invented model of how our minds work. No one can see seeing, concepts, and what-is-seen, and then visually compare them. Even a meager awareness of the history of psychologizing, from Parmenides to Piaget, will help to appreciate why Einstein was perfectly right to call the problem of analyzing analysis and other psychological acts "much more difficult" than doing physics. The reason is simple. By the time we are five, we are used to thinking about bodies, but not at all used to thinking about thinking.

Suppose then that we apply the idea that our mind must use models even when we try to understand thoughts about our mind. It is false, of course, to think our mind uses models. Persons use them. You use them, I use them, he and she use them. Even to (falsely) think memory-images are models supporting our theories is pretending. But memory-images and thoughts, even false or let's-pretend thoughts, are real.

Everyone who invents a model for exploring the jungle that is human experience must begin with our everyday thinking about sensing. We begin with our everyday idea that we are or have a body and that it has five parts related to sensing. We use eyes to see the shapes and colors of things, ears to hear the sounds things make, hands to feel eyelids with, and so on. Each (we realize later) is defined in relation to the others. Just so, the congenitally blind and deaf hear 'words' the way we heard 'science' and 'Santa,' and use the idea of their body as a basis to create their own models.

But there is more than sensing. We see, but we can also puzzle over what seeing is. We hear what others tell us, but we can also begin asking

249

I See Colors

for definitions. We take medicine that is bitter and eat food that is sweet, but we can also wonder why we can tell the difference between bitter and sweet. We know people prefer having sex to remembering it, but prefer remembering headaches to having them, but we can also stop and ask how to have more pleasure and less pain. We do more than sense.

Socrates and Plato concluded that a two-part model is needed. The bodily eyes, ears, etc., are used for the seeing, hearing, smelling, tasting, and feeling mentally grouped under "senses," but something else named "the psyche" is needed for the 'more' that we do. That two-part model has been the foundation for theories about human knowing ever since. It was the foundation Descartes constructed his *First Meditation* around, the base Kant built his new model on, the distinction Einstein based his theorizing about theories in physics on.

Plato took the next step. In Book IV of his *Republic,* he added further detail to the psyche or soul half of the model. In the same way that the body has five organs for five types of sense acts, Plato wrote that the psyche has three 'organs.' He showed how the various conscious acts we know we perform are best imagined as coming from different parts of the psyche or soul. There is a part involved when we desire things. There is another part involved when we resist certain desires. And there is a part involved when we reason about such things as worthy vs. unworthy desires.

When Descartes composed his *Meditations,* he radically changed the model. He took all power of knowing away from the body, including the brain. That is why he held that animals are not conscious. As for the eyes, ears, etc., they do nothing but filter heterogeneous stimuli (light, sound, particles, pressures, etc.) from the outside environment, send homogeneous currents through nerve-tubules to the brain, at which point ideas — if nothing else — are experienced in the mind. To describe the activity of the self or mind requires more than just three concepts.

> But what then am I? A thing that thinks. What is a thing
> that thinks? It is a thing that doubts, understands, [conceives],
> affirms, denies, wills, refuses, which also imagines and feels.

The Wonderful Myth Called Science

(R. Descartes, *Meditations*, trans. E. Haldane & R. Ross, Vol. I, p.153)

Between Plato and today, endless new concepts of 'mental acts' have been created to capture subtle facets of our endlessly-faceted thinking vis-à-vis its various objects, including sense-data and images, as well as between different complete thoughts (propositions) themselves. The list is a long one: analysis, synthesis, abstraction, apprehension, composition, division, judgment, affirmation, negation, reminiscence, association, connotation, denotation, abduction, reflection, introspection, inference, creation, etc., etc., etc. Where some writers distinguish, others reduce. Can there be any doubt, however, that — except for an innovator who creates a new concept and a new name for it — most of us acquire our own ideas of all those 'mental acts' from words, words, and more words? Mostly read words?

Question: Look at the list and ask, "How can I be certain which of those acts I have been performing since I began reading this book?"

Answer: A better question would be, "What acts does reading require?" The answer here is that (i) you have been understanding thoughts. (Can't you be sure of that?) (ii) You have been relying on an inner model of the world (have you been reading up on the moon or at the bottom of the sea?) which includes your body. (Have you been using your nose to read?) And (iii) you've been seeing this and all the other things that seem to be out here on what you'd ordinarily think is a page. (Right?)

But you haven't been seeing any acts out here on the page. No models here, either. Inspired by Kant who insisted that our mind is preprogrammed to create the ideas needed to 'put order' into the sense-data and images, Einstein argued that, in the same way that we create concepts of physical things, we must create our idea about the difference between sensing impressions and creating concepts.

A few more remarks of a general nature concerning concepts and [also] concerning the insinuation that a concept— for example that of the real—is something metaphysical (and

251

therefore to be rejected). A basic conceptual distinction, which is a necessary prerequisite of scientific and pre-scientific thinking, is the distinction between "sense-impressions" (and the recollection of such) on the one hand and mere ideas on the other.

> There is no such thing as a conceptual definition of this distinction (aside from circular definitions, i.e., of such as make a hidden use of the object to be defined). Nor can it be maintained that at the base of this distinction there is a type of evidence, such as underlies, for example, the distinction between red and blue. Yet, one needs this distinction in order to be able to overcome solipsism. (A. Einstein, "Reply to Criticisms," p. 673.)

We do not create our sensations. We do not create colors, such as red and blue. We may create our concepts or ideas of colors, but the colors are 'effects in us.'

What, then, is seeing? The answer must include what it is not. It is not hearing, smelling, tasting, tactile feeling. It must correlate with colors, just as hearing does with sounds, smelling with odors, tasting with flavors, feeling with heat or coolness. The third leg of this argument will focus on the question, "What are colors?" Once that question is answered correctly, it will be clear why "seeing" is best analyzed as 'being directly aware of' or 'visually experiencing' real colors.

D. THE COLORS I SEE

What needs to be explained?

What is the *explicandum*? That's the Latin derivative some experts use as shorthand for "What needs to be explained?" Here, it is that colors are what-we-see. Whoever reads about what-we-see knows what "we" and "see" mean. Whoever reads about colors must know what "color" means.

The Wonderful Myth Called Science

This may seem obvious. But our belief about seeing colors is as distinct from seeing colors as conceptual thinking is distinct from any sense experience. (Einstein's postulate.) Being blindfolded prevents us from seeing colors, but being blindfolded doesn't affect our beliefs about seeing colors one little bit! We could see when we were born. But we could not think such complex thoughts as the ones expressed by these sentences till we were older. By age two? At least by age five or six.

That is such an important principle that it demands thoughtful reflection. There are many things five-year-olds do not know. But knowing what "Peekaboo, I see you" means is not one of them. They may be mistaken about what they see, as they are about Santa Claus coming to town, but they know what "I see you" means. To imagine college students do not know the difference between seeing colors and hearing sounds until they study a psychology text book is equivalent to imagining that they do not know the difference between the blind and the deaf till then. As noted earlier, no one goes to the doctor to learn what "pain" means or to find out if they have one.

Obviously, certain things are being presupposed here. If you can see 'this page,' then you can see. If you can understand that sentence, then you know what "seeing" and "page" mean. But, though you do see, and though what you see is color, the color has no direct connection to any page. "This page" refers to a fiction of your imagination. The page does not exist, but the colors you see do. If you understand that sentence, then it is an additional piece of evidence that you know what "seeing" and "page" mean. This final section of Chapter V will explain or lay out in more detail the case for Russell's 'masterful' conclusion that we sense effects, e.g., colors, in us, which means that the naïve-realist belief that we see such physical things as green grass — or print-covered white pages in a book — is an illusion.

Effects that are in us (me).

The colors we see are in us. Colors and sounds exist. We can be certain they exist, at least during the time we are experiencing them. But that they are in us comes as a complete shock. It means that the starry

253

I See Colors

night sky is not what we see. The tiny white pinpoints we mistake for stars billions of miles away are in a total visual field or TVF that's in us. That visual field is annihilated the instant we 'feel' our eyelids pull down and block the unseen light that came from so far away.

"In us." Or, more helpfully, "in me." The phrase must be pondered. Savored. Regardless of the excitement or terror that it evokes. Theoretically, it is obvious that "what is in me" and "what is outside of me" mean opposite things in numerous everyday contexts. My headaches are in me, the aspirin before I take it is outside of me, and the aspirin when I swallow it is in me. After they have learned how to use "I" and "you," children know how to switch from "in me" to "in you." Hence, your beliefs about what you see are in you.

Just as your feelings of love are in you, your seeing is in you. The shock, once again, is that what you see is also in you, unlike the person you love who is not in you. Recall what being newly in love with someone is like. When we are out of their presence, they are still 'present.' Our thoughts frequently turn to them. We may be otherwise occupied when suddenly something we see or hear makes us think of them. But the thoughts, memories, images, and feelings we have are in us, whereas the person is not.

But where in me are the effects Russell wrote about? Where?! And how can I nail down the answer? Russell wrote that Watson never saw a rat (or Little Albert), only "events in himself." Russell added that no neurologist ever sees a brain, only "percepts which he himself has." So, where are these events or percepts? And how can I ever be certain I have not done what Descartes was afraid he might do with respect to his self, namely, mis-take one thing for another? Such mis-taking is analogous to mis-taking the forged painting here for the authentic one that's elsewhere.

". . . and physics, if true, shows that naïve realism is false." Shows to whom?

Nowhere, perhaps, does our intuitive sense of logical consistency display itself with such glaring clarity as it does when we are first confronted with the claim that those are not stars we see in the night sky but effects in us. That the pain is in us, not in the needle jabbing us, that the tickle is in us, not in the feather brushed under our nose, even that the actually-heard sound is in us, not in the unvisited forest, are easier to 'take' than the claim that the 'wide-screen' colors which seem so obviously distant from us are not distant, because they are effects in us.

That intuitive sense of logical consistency almost forces all of us to look for ways to 'get around' whatever alleged proofs are used to show that what we so obviously see are not outside of us, in the physical things that are. Among the dodges are redefinitions of "color," redefinitions of "see," and redefinitions of "inside the mind." However, since current illusions about science and the scientific method include the dogma that scientific theories must either be verified or falsified by observation, it is only reasonable to ask how such a dogma can itself be verified or falsified by visual observation or seeing.

Inasmuch as most of what you have taken to be science was learned by hearing or reading words, and since you are now seeing while you are reading, it is convenient to continue using the seeing you are doing right now as the test for what you are, for what your seeing is, and finally for what it is that you see.

But the physics Russell referred to will never show you that naïve realism is false if you are determined not to allow it to do so.

"In the mind": another stipulative definition.

The thesis here is that, like images of colors and like thoughts about colors, the colors themselves that we have images of and thoughts about are 'in our mind.' Colors are sense-data, one of the three types of real things 'making up' our seemingly seamless stream of consciousness[*]. All three types are immaterial.

What does it mean to say they are "in the mind"? The best way to explain how "in the mind" is used in this text is as follows. (i) It is not used to mean "non-existent" or "purely imaginary," as when we use it to say that Mordor exists only in the minds of people who have read *Lord of the Rings* or seen the movie(s) or to tell a paranoid person that the FBI plot against him is all in his mind. (ii) It is not used to mean "above your shoulders, in your head." Many people today use "mind" and "brain" interchangeably, which means that when they say something is 'in the mind,' they think it is somewhere in their head. (iii) The predominant sense of the term as it is used here is thus partly negative. This stipulative definition puts a fence around the first two ideas and says it is not synonymous with either of them. As for the positive sense, it is an assertion that sense-data, memory-images of them, and ongoing thoughts really exist as such and are different from imaginary and physical things, such as the brain.

The phrase "in the mind" can obviously be used, therefore, in more than one way. In each case, the mindset-context of the phrase-user determines what thought it is meant[†] or intended to convey. Sense-data are more than any and all physical things.

-----

[*] "Consciousness" reifies the adjective "I am conscious."

[†] "Meaning" reifies the verb

The Wonderful Myth Called Science

The sense that there's more: Key to learning more and/or new insights.

Part of Einstein's favorite answer to "How do we learn more?" would be "We create new concepts or ideas." Recall the way we acquire the ideas of Santa Claus and science. We first must have the inner mapped model of the outer world referred to in Chapter IV, then must acquire habits of associating certain sounds with major items in that mapped model (mom, dad, kitty, etc.), become accustomed to some new 'word,' and in time create the additional concept to go with it.

The major problem in learning comes from the ambiguity of words. The trouble, however, is not in the words, since they do not exist. It is in thoughts. What one word-user is thinking may not be what we think they are thinking. A materialist can say "Color is all in our mind" and think about what's inside the skull, whereas an idealist may say the same thing and mean something wholly different. The only solution is to learn, not just how to use a guess-list dictionary, but to learn enough about the radically contradictory worldviews used by writers who use such non-words as "word," "name," "science," and here "color."

Learning a radically new worldview-philosophy later on can only be done by using our first, common-sense worldview. For instance, to learn how a materialist or idealist thinks, we must take that original mapped model after its major items are 'tagged' with 'names,' then use it and them to create — or remove! — concepts. Or, here, to understand the new thoughts about what exists and what doesn't.

That is why the positive way to define, describe, or try to convey the thought that color is 'in our mind' is to say "Color as such is more than, over and beyond, in addition to, different from physical bodies, whether large or small, including (if it exists) the brain." In the same way that thought is more than sense experience and Neptune more than Uranus, thought plus sense experience plus imagery are more than anything physical. "Physical things" is shorthand for the stars, the local sun/star, the planets, vegetation, animals, and human bodies. "Immaterial things" here is shorthand for persons, thoughts, sense-data, and memory-images of sense-data.

Today's scientific psychology: help or hindrance?

An earlier chapter very briefly touched on the chaos that exists among the individuals whom we mentally group under the label "scientific psychologists." The point was made that most of its practitioners believe genuinely scientific psychology must ultimately be based on observation of others, whether the others are maze-running rats, computer-poking chimps, unborn fetuses, newborn babies, children tested for ADHD, questionnaire-taking college students, etc. The method, as Ulric Neisser noted, was "Avoid all subjective observer-contamination of the objective outcome." No introspection, please.

One result was that psychologists writing about seeing could not begin by reflecting on what they themselves experienced, so as to feel certain they knew what "see" means. Even now most use a dodge. They combine seeing and thinking into a hybrid named "perception." This permits them to write as if sensing is pure physics, chemistry, and physiology.

Consider again the experience that Dennis the Menace lacks, namely, falling in love with a specific person. It is obvious that what lovers remember is in them and what they feel is in them. What makes them happy or sad is not the beloved as such, who exists independently-of-them, but their beliefs about the beloved, beliefs which also are in them. Unwittingly-betrayed lovers who are happy but deceived, irrationally-jealous lovers who are miserable and mistaken are proof of that. If Dennis does not know the meaning of "being in love," and he doesn't (he is also as mythical as Santa), then no psychologist will ever know what the feelings named "love," "jealousy," "happiness," or "misery" are like who has never felt them. Like blind people who think about an unknown-X called color, such psychologists might use the 'symbols' for seeing and color, but they'll never know what they are talking — or rather, thinking — about.* No introspection? Please!

Yet, ironically, today's psychology texts are excellent tools or instruments for discovering how true Russell's Einstein-approved formula is. "Naïve realism leads to physics, and physics, if true, shows that naïve realism is false."

At each step, ask "Is this the color I am seeing?"

Today's 'scientific' texts help by filling in minute details of the 'body' half of Descartes' revolutionary model. If there is a 'body' half, i.e., if there is a physical world at all, then every reader can use those texts to do what Einstein said, create the theoretical fictions used to formulate laws of physics and physiology, thereby preparing themselves to test the model against their own stream of private consciousness.

Earlier pages outlined the most important details of the physics-physiology of the senses: We come into this world belief-less. We — at least our bodies — are bombarded immediately by heterogeneous stimuli — light, sound, odors, heat, etc., coming from the environment where they are jumbled together. First, they are filtered by our five sense organs. Eyes pick out light, ears filter out sound, skin responds to kinetic energy, resistance, and so on. The different or heterogeneous stimuli,

---

* 'Scientific psychologists' often write knowingly about such things as seeing, which — were they actually relying on the method they profess and not on what they learn by personal experience — they would never suspect of existing and never write about.

259

after being filtered, trigger non-differentiated or homogeneous electro-chemical chain-reactions conveniently lumped together under "nerve-impulses." Only these homogeneous nerve impulses, traveling via afferent neurons, reach the brain, at which point they correlate with.....

The claim here is that the homogeneous brain processes correlate with the observed heterogeneous 'effects' in our mind, namely, colors, sounds, odors, etc.

Since the sense that readers rely on the most is seeing, the effect in us is the spread-out field of colors, the visual component of our virtual-reality life movie. The relevant parts of the physical-world model are (i) light-radiating bodies such as the stars, including our local sun/star, the traveling light itself, (ii) the camera-like eyes that focus scattered light into coherent patterns popularly referred to as retinal images, (iii) chains of electro-chemical reactions along the optic nerves, and similar electro-chemical reactions in various parts of the brain.

The method for precisely locating the things every normal person sees, whether star-gazer or reader, is plain: compare the created concept of each part of the current model with what we see, visually-observe, or experience-via-sight. (The thought's the thing, not the black figures.)

Start with stars, which all of us believe are the most distant things we can see. As explained previously, no one sees stars. They may no longer even exist by the time much-travelled light arrives to create effects in our earth-bound eyes. Besides, for all the reasons Galileo, Descartes, Locke, Newton, Berkeley, and Leibniz pulled together, no physical bodies possess color as such.

Some thinkers have proposed that we see the light-patterns hitting our eyes. If this were true, we would see two visual fields, one per retina, and both would be upside down. In addition, light itself is colorless photons, some of which are closer together and others farther apart as they travel. Newton warned against identifying light with colors as such.

> If at any time I speak of light and rays as coloured or
> endued with colour, I would be understood to speak not

philosophically and properly, but grossly, and according to such conceptions as vulgar [naïve] people in seeing all these experiments would be apt to frame. For the rays to speak properly are not coloured. In them there is nothing else than a certain power and disposition to stir up a sensation of this or that colour. (I. Newton, *Opticks*)

I referred earlier to d'Abro's theory that no one not mathematically advanced can ever discover flaws in Einstein's or Bohr's far-out theories. He was wrong about that, but he helped to distinguish light from 'stirred up sensations' of color. Newton knew prisms break sunbeams into spectra of velvety colors, but not why. ("Red light" means light that 'stirs up' the sensation of red color.)

As soon, however, as Fresnel discovered the vibratory nature of light, red light was found to differ from green light owing to its slower rate of vibration; prevision then became attainable. It was possible to anticipate that were we to approach a red lamp with sufficient speed it would appear green, that with greater speed it would appear violet, and that with still greater speed it would become invisible. Likewise, were we to recede from the light with sufficient speed it would also cease to be visible. This was the celebrated Doppler-Fizeau effect. (A. d'Abro, *op. cit.*, p.389)

Anyone who directs a laser pointer at the ceiling can testify to the fact that he or she sees a precisely located colored dot, but no light beam travelling from the pointer to the ceiling. Physicists see red and blue colors, they guess light. As children, they had ideas of red roses and blue violets, they heard about wavelengths and frequencies much later. Hence, Einstein's comment, cited at the beginning of this chapter, about the difference between the merely thought-about distinction between sense experience and thought and the actually-experienced difference between colors:

Nor can it be maintained that at the base of this distinction there is a type of evidence, such as underlies, for example, the

261

distinction between red and blue. Yet, one needs this distinction . . .

A third group of thinkers believe that what we sense are the colorless nerve-impulses in our brains. Not only are young students no longer taught that their mind — their self — is a radically different thing from their never-experienced brain. They are positively encouraged to use the increasingly-popular adage, "The mind is what the brain does," as a philosophy-avoiding mantra.[*]

Why is it an error to believe that what we sense are the colorless nerve-impulses in our brains? There are several reasons. First and foremost is the fact that brains as such do not exist. More than a century ago, William James called the created-concept of unified brains "a fiction of popular speech." Research indicates that neurons are not continuous with each other, but are on the contrary separated by gaps called "synapses" which means "connections" (an oxymoron if ever there was one!). Rutherford's research indicates that not even neurons exist, only bodies that are subatomic in size and individually distinct from each other.

The second reason, only a tiny bit less important (if at all), is that, once we begin our quest to locate the 'effects in us,' we must not forget that we already know what the effects in us are. Colors! The real question is not "What are the effects?" The question is "Where in us are they?" The reasoning so far has shown that the effects that we experience come at the end of the before-and-after chain of apparent cause-effect

---

*Many materialist philosophers use that same adage. Other philosophers, however, have invented complex vocabularies in order to avoid taking sides on the "What exists?" question.

links that precede our experience of colors. Nothing in the brain is in any way even remotely like the colors we experience.

In other words, once it is clear that "Is this the color you are seeing?" is the question to be asked about each separate item in the cause-effect chain, it becomes equally clear that, at each point along the line, the answer is "No, something else must be the effect I am looking for. The colors are more than any of the physical things thus far proposed, if . . ."

If physical things even exist!

It's a law of nature . . .

It's a law of nature that long-standing habits are hard to change. Naïve-realist beliefs are habits of thought. They are much longer-standing than the habit of measuring distance in feet, yards, and miles, instead of meters and kilometers, *yet how difficult it is to change that habit.* Naïve-realist thought-habits are like the roots of a giant tree that grow outward and wrap themselves around other basic beliefs that seem too true to be mistaken. That is the first reason for many people's defiant refusal to take Descartes, Locke, Berkeley, Hume, Kant, and the other modern psychology greats seriously.

It's a law of nature that long-standing habits are hard to change. In modern times, most of us learn at some point to re-define "color." It may be redefined as light-reflecting and light-absorbing properties of physical things. Or as light itself, light-frequencies or light-waves. Such redefinitions are on a par with Spinoza's and Einstein's switch from using "God" for the traditional idea of a super-intelligent, omni-competent creator, to using it for the idea of an unintelligent non-creator. Even when this is pointed out, it is hard to realize at first why it is a redefinition. After all, if research has shown that water is really H^2O, why is it wrong to think that research has shown the real nature of color to be light? One easy answer is that water, hydrogen, and oxygen do not exist. An easier answer is that sighted people see color, but no one has ever seen light. No one ever sees, period, till nerve impulses, not light, arrive inside the skull's pitch-dark recesses.

I See Colors

But because the temptation to equate color as such with light is as hard to resist as it is to surrender the 'socially-approved' myth called science, a few further thoughts may help. You can look at this book for as many millennia as you wish, but you will never see any light bouncing from this page into your eyes. If you saw the light, you wouldn't see the black figures against a white ground (as the Gestaltists describe it). Why? Because, if you saw the light, it would block out the page. It would be like driving through fog and seeing only the lit-up water droplets caught in the headlight beams. When we think we are seeing a beam of sunlight coming into the living room, it is not light but the floating dust we see. (If, that is, we see anything not inside our skull.) But the best place to begin realizing that we do not see light as such is to look at the moon on a clear night and to realize that the entire night sky is flooded with sunlight, everywhere except for the long-tunnel shadow created when the other, lit-up side of the earth intercepts some of the sunlight. (We call it an eclipse when the moon moves into that shadow.) This could go on and on, but space and time are not infinite.

It's also a law of nature that, on such a basic issue as "What do readers see?", a radically anti-naïve-realist decision will affect one's belief-decisions on other basic issues, e.g., "What did Einstein, Bohr, and others mean by 'observe'?"

The problem of perception has been solved, dozens of times!

"What do I see?" Any well-stocked library will offer resources for anyone who has time to study the vast range of answers given to that question. The contention in this book is that no one will ever discover an answer different from the ones already proposed. Here are a few of them.

1. I see a page in a book. (G. Ryle and other 'analytic philosophy' naturalists)

2. I see the colors (qualities) of the page and the ink. (Aristotle)

3. I see the light-absorbing and light-reflecting properties of the page and the ink.

4. I see a colored image in each eye, reflections of the outer colored things.

5. I see light's various frequencies and wave-lengths. (2002 general psychological text)

6. Our eyes detect light waves. (2003 'Psychological Science' textbook)

7. Seeing takes place in the brain, not the eye. (2002 general psychological text)

8. My brain (mis)interprets nerve impulses as color. (Numerous neuroscientists)

9. Color is something the brain creates. (2002 general psychological text[*])

10. Color is a useless by-product of brain processes. (Huxley)

11. Color is a colorless idea in the mind 'of' a quality outside the mind. (Locke)

12. Colors are immaterial things I am directly aware of. (Berkeley and here)

13. I see nothing. (Congenitally blind people)

14. I see nothing. (Armstrong and all who re-define "seeing" as thinking.)

---

* If the psychology texts are scientific, shouldn't they agree on a subject so central to theories about science?)

I See Colors

Who solved it first? George Berkeley.

Descartes set the stage. He first recognized we do not sense things outside our mind, only inside effects. But he oscillated between ideas 'of' sensed effects and sensed effects as such, a huge difference. An idea of pain doesn't hurt, but pain does!

> As to other things such as light, colours, sounds, scents, tastes, heat, cold and other tactile qualities, they are thought by me with so much obscurity and confusion that I do not even know if they are true or false, i.e., whether the ideas which I form of these qualities are actually the ideas of real objects or not [or whether they only represent chimeras which cannot exist in fact]. . . .

> Nature also teaches me by these sensations of pain, hunger, thirst, etc., that I am not only lodged in my body as a pilot in a vessel, but that I am very closely united to it, and so to speak so intermingled with it that I seem to compose with it one whole. For if that were not the case, when my body is hurt, I, who am merely a thinking thing, should not feel pain, for I should perceive this wound by the understanding only, just as the sailor perceives by sight when something is damaged in his vessel. (R. Descartes, *Meditations* III and VI, trans. Haldane & Ross, pp.164, 192)

Feeling pain is not having a thought about pain. If sensed color and felt pain are both effects, and both caused by the brain, then seeing the color 'of' the blue sky is not just thinking, either. Real hallucinating is not just thinking. Thinking can be done anytime!

Where Descartes flip-flopped on sensing, Locke set forth a new, clear model. But in doing so, he adopted the flop-side of Descartes' thinking. In Book II of his *An Essay Concerning Human Understanding*, he described sensing as the mind knowing certain ideas caused by brain-activity caused by eyes, ears, etc. Those certain ideas are ideas of various colors, sounds, odors, etc. Seeing, as opposed to hearing, etc., is an act of the mind becoming aware of shapeless, colorless ideas, chiefly of color and shape.

Berkeley corrected Locke. (Care must be taken while reading Berkeley; unfortunately, he followed Locke by calling colors, sounds, etc., "ideas.") He took a huge step forward when he realized the monumental difference between colorless ideas OF color, on the one hand, and color AS SUCH, on the other. It parallels the difference between odorless ideas OF odor and odors AS SUCH, of painless ideas OF pain, and pain AS SUCH.

The bottom line for Berkeley and here is that our mind — we — are aware of or experience colors that are effects in us. We can only guess at or infer any unsensed causes, whether stars, the brain, or anything else.

Hume's bombshell: "How can you have an idea of a never-sensed cause?"

Previous chapters have already referred to Hume who woke Kant from his dogmatic slumber. Kant had already absorbed Descartes, Locke, and Berkeley. What shook him from the top of his head to the tips of his toes was Hume's attack on our everyday cause-effect thinking.

Hume began with naïve realism. We may see billiard ball A strike billiard ball B. But, if we take care to not theory-ladenly project what we think onto what we sense, it becomes clear that we do not sense any causal force or power being transferred from A to B. Billiard ball A makes contact with billiard ball B; B moves. We see two events, one after the other. In the same way that Freud suggested we mistakenly 'project' onto the cigar what isn't there, so we mistakenly 'project' force onto what we see. It is comparable to projecting onto the paper and coins our idea of $$$, and onto the movie screen colors the idea of Snow White who isn't there.

What happens when we apply Hume's reasoning to 'modern physics?' We put a pan of cool water on the stove, turn the knob, flames appear, the water heats. We say the flaming gas causes the heat. 'Scientists' guess that unseen photons of energy are emitted from the oxidizing gas and cause greater Brownian motion in unseen molecules. Further reading shows us that's guessing, too. Even if unseen electrons exist and do jump from a higher to a lower orbit and photons emerge, no one can begin to say why. An atomic force? Electromagnetic energy? Those concepts go into the same 'fiction' class with never-seen gravity.

Hume next applied the conclusion Berkeley had drawn from the premises discovered by Locke and Descartes. Just as we never see tables, only sense-data in our mind, we never see billiard balls, either. Like movie-goers, we experience only a succession of sense-impressions. During the day, as during the movie, we do not even notice that we never see 'the same thing' twice. Each new reappearance of Snow White is similar to the earlier thing we only remember. Each new glimpse of the table is similar to the glimpse we now only remember. Snow White herself never appears. Of course, the table never appears. Billiard balls and their motions are never seen, either.

Hence Hume's famous deduction: We have no ideas that are not copies of what we sense, we sense nothing but series of before-and-after impressions, therefore the idea that we see one body 'causing' motion in another should always be analyzed into the thought, "Event-B types are always conjoined to antecedent Event-A types." If A is called "cause" and B is called "effect," both the 'cause' and the 'effect' are sensed, observed, experienced. It is illegitimate, Hume insisted, to argue, as Descartes, Locke, and Berkeley did, that what we sense are effects of a never-sensed cause, whether it's an unsensed brain or an unsensed creator. It is illegitimate to build a theoretical, deductive bridge from private events inside the mind to never-experienced causes outside the mind. That includes the (imaginary) brain.

What Hume actually did: created another re-defining ambiguity.

The truth about Hume's cause-effect analysis is that he dismissed our common-sense idea and substituted the before-and-after, 'law of nature' idea. Laws of nature are shorthand descriptions of what happens, not why they happen. Not content with this, Hume went further and postulated that the event-event laws of nature are unvarying, exception less, iron-clad, before-and-after, hence lawful descriptions. This switch from cause-effect to event-event thinking about nature is crucial for understanding Einstein and nearly all modern thinkers who refer to quantum indeterminacy as 'the downfall of causality.' To such thinkers, adopting quantum indeterminacy must seem akin to believing in routine miracles, i.e., exceptions to iron-clad 'laws of nature.'

However, contrary to Hume's postulates, our everyday-thinking, common-sense idea is that, when something happens, something makes it happen, whether the latter is seen or unseen. The common-sense 'causal sense' is not restricted to sensed causes. Agree or not, everyone can understand what is meant by saying that forces, particles, orbitals, dimensions, and other 'entities' are causes. Agree or not, everyone can understand what is meant by saying that never-seen light and brains are causing colors, sounds, and other sensory phenomena.

Few issues are as crucial as the issue of causality or cauation. And few protests against the theorists' inconsistencies in dealing with it are as noteworthy as the protest James recorded in his *Principles of Psychology:*

> As in the night all cats are gray, so in the darkness of metaphysics all causes are obscure. But one has no right to pull the pall over the psychic half of the subject only, as the automatists do, and to say that that causation is unintelligible, whilst in the same breath one dogmatizes about material causation as if Hume, Kant, and Lotze had never been born. One cannot thus blow hot and cold. One must be impartially naif or impartially critical. (W. James, *Principles of Psychology*, Vol. I, p. 137)

The bottom line was stated succinctly by Wittgenstein who wrote, "The whole modern conception of the world is founded on the illusion that the so-called laws of nature are the explanations of natural phenomena." (L. Wittgenstein, *Tractatus Logico-Philosophicus*, #6.371) Here, of course, it is those phenomena themselves we are trying to get clear about.

I See Colors

Hume's Magic Moment and our vast inner 'world.'

Hume once described a miraculous experience, a kind of 'Magic Moment' insight into our vast inner 'world.' It appears near the end of Vol. I, Bk. I, Pt. I, sec. vii of his *Treatise of Human Nature*:

> And, indeed, if we consider the common progress of the thought, either in reflection or conversation, we shall find great reason to be satisfied in this particular. Nothing is more admirable than the readiness with which the imagination suggests its ideas, and presents them at the very instant in which they become necessary or useful. The fancy runs from one end of the universe to the other, in collecting those ideas which belong to any subject. One would think the whole intellectual world of ideas was at once subjected to our view, and that we did nothing but pick out such as were most proper for our purpose. There may not, however, be any present, beside those very ideas, that are thus collected by a kind of magical faculty in the soul, which, though it be always most perfect in the greatest geniuses, and is properly what we call genius, is however inexplicable by the utmost efforts of human understanding. (*Op. cit.*, p.31)

No newborn infant has a 'world of ideas.' By age five, the normal child, including every aborigine child, does have a vast 'intellectual world of ideas,' but not the faintest idea that it has it. Many people have a vast knowledge of the physical world got from various sources, mostly reading, but not the faintest idea that what they think of as the vast physical world out there is a vast 'intellectual world of ideas' in their mind, a model of the physical world.

The quintalist thesis is that everyone's daily life involves something comparable to watching a movie. Daily life involves a flow of always-new sensations, images, and thoughts. Paradoxically, out of the flow we slowly build up a stable inner model of what we believe is the outer world. To notice our mind's vast, experience-acquired stable model, it is essential to distinguish it from the sense-data which are there-and-gone as fast as the changing colors on a movie-screen and the loudspeaker sounds that each last no longer than a there-and-gone chord from Beethoven's Ninth.

The quintalist model incorporates Hume's idea of imagery and association. An in-pouring symphony of colors, sounds, odors, tastes, warmth, coolness, pleasures, and pains is 'there' the moment we emerge into the delivery room. At first, no backdrop of criss-crossing, associated memory-mages 'of' previously experienced colors, sounds, odors, etc., is also 'there,' waiting to be specifically activated by each moment's new sensations, but in time there will be.

Hume's great discovery was that, besides the spatial inner 'world' model that we constantly add details to, we have — as memory-base for 'laws of nature' — short memory-clips or image-videos of countless temporal, before-after scenarios: of the sun's daily transit across the sky from east to west; of an acorn's growing into a seedling, sapling, and lastly a tall oak that sheds leaves each autumn and grows a new coat of them each spring; hens laying gooey-inside eggs from which emerge fantastically complex peeps; trips to restaurants where we eat, pay, and leave (plus all the details Roger Schank includes in his scripts); the sequential aimings and shootings that 'make up' a game of billiards; etc.

In time each of those oft-observed event-sequences is 'fixed' in memory and associated with remembered sounds associated — for readers — with remembered ink marks, e.g., "hens laying gooey-inside eggs." As readers' eyes receive new light-input from ink marks, the resulting sense-data spark a massive, lightning-fast, and orderly retrieval of imagery-plus-thought. (Did you notice?)

> The fancy runs from one end of the universe to the other, in collecting those ideas which belong to any subject. One would think the whole intellectual world of ideas was at once subjected to our view, and that we did nothing but pick out such as were most proper for our purpose. (*op. cit.*, p.31)

I See Colors

The ultimate "Why?"

The ultimate 'causal' question is this. How do all of the ideas correlated with the 70, 000 or so words, which we normal adults learn, get into our inner Big-Picture 'worlds' of ideas? The thesis here is that many or most of those ideas are created *in conjunction with reading*. Whatever we learn from writers we have never met, especially if they are dead, must be coming — it seems — from their words. That is how the experts, too, learn most of what they learn. *The Science Times* for 9-24-02 described the ten "Most Beautiful Experiments" in the history of physics. Kepler and Einstein did none of them. The bulk of their 'scientific observing' was seeing only what other readers see.

Theses can be tested by experiment. Even purely 'pragmatic,' i.e., false but useful, theories 'work' if they accurately predict future sensations. Do readers' ideas really come from what they/we see? Sherlock's advice would be to eliminate what's impossible, and whatever is left must be the truth, however improbable it seems at first.

Scanning black figures against a white ground can be eliminated as an answer to the question. At first, it seems that we do get ideas by scanning row after row of tiny, shaped and arranged, black marks. The marks are like subtitles on a movie screen. As you scan them, they pass through the very center of your ever-changing visual field. Test this first hypothesis. i— Close your eyes and notice that the visual movie instantly disappears, the backdrop of associated imagery runs less predictably, and the flow of reading-directed thought stops. ii—Open your eyes once more, run your eyes across the tiny black ciphers, and the directed flow resumes. (This is what happens for me; on July 4, 2005, I am guessing it will do it someday for you, too.) But . . .

But it's an illusion. Long ago, Plato hypothesized that such learning is recalling, not looking or, in this case, scanning. St. Augustine thoroughly tested the thesis in *De Magistro* (On the Teacher). And John Locke nailed down the truth of Plato's hypothesis.

No new simple idea can be conveyed by traditional revelation.—First then I say, that no man inspired by God can

The Wonderful Myth Called Science

by any revelation communicate to others any new simple ideas, which they had not before from sensation or reflection. For whatsoever impressions he himself may have from the immediate hand of God, this revelation, if it be of new simple ideas, cannot be conveyed to another, either by words, or any other signs. Because words, by their immediate operation on us, cause no other ideas, but of their natural sounds: and it's by the custom of using them for signs, that they excite and revive in our minds latent ideas; but yet only such ideas as were there before. For words seen or heard, recall to our thoughts those ideas only, which to us they have been wont to be signs of; but cannot introduce any perfectly new, and formerly unknown simple ideas. (J. Locke, *Essay Concerning Human Understanding*, Bk. IV, Ch. XVIII, sec.3)

". . . words, by their immediate operation on us, cause no other ideas, but of their natural sounds." Had his attention not been fixated on the Bible, Locke might have generalized by adding, "Any additional ideas you are getting while reading my words are from some other source, e.g., from the 'whole world of intellectual ideas' already stored in memory. But where is that? Certainly not in your never-sensed brain.

First, according to Hume's 'copy' theory, we cannot have an idea of a brain. No one has ever seen one, hence no one can have an image or faint copy of one. Secondly, even if brains existed, Hume would have to deny that they cause any part of the stream of consciousness — namely, sense-data (impressions), images (copies), or complete thoughts — since that would make a never-observed, only-postulated brain a cause in the common-sense sense of "cause." Whoever accepts Hume's analysis of causality and wishes to be logically consistent will have no remaining alternative than his new, sequence-of-impressions or laws-of-nature redefinition of "cause." Finally, every appeal to the brain must be ruled out for the simple reason that Rutherford's discovery shows that brains as such do not exist, only subatoms dancing in empty space. (It is an interesting exercise to ask, "Is it my brain that's telling me it doesn't exist?")

273

I See Colors

What about the laws of nature? Can they help? No. Even if 'laws of nature' did exist, we would have to rule them out as the bottom line for all of this.

> In practice we come rather quickly to laws which cannot be explained further. Laws about atomic structure are typical of such laws. Laws of psycho-physical correlation are another example. Why do I have a certain color-sensation which I call red, indescribable but qualitatively different from all others, when light within a certain range of wavelength impinges upon my retina, and another indescribably different sensation which I call yellow when rays of another wave-length strike the retina? That this wave-length is correlated with this visual experience seems to be sheer 'brute fact' —a law which cannot be explained in terms of anything more ultimate than itself. . . . Like so many others, this point may seem logically compelling but psychologically unsatisfying. Having heard the above argument, one may still feel inclined to ask, 'Why are the basic uniformities of the universe the way they are, and not some other way? Why should we have just these laws rather than other ones? I want an explanation of why they are as they are.' I must confess here, as an autobiographical remark, that I cannot help sharing this feeling. I want to ask why the laws of nature, being contingent, are as they are, even though I cannot conceive of what an explanation of this would be like, and even though by my own argument above the request for such an explanation is self-contradictory. (J. Hospers, "What is Explanation?", in *Essays in Conceptual Analysis*, ed. Antony Flew, 1956)

Which leaves us with a question. Is it really that impossible to conceive what another explanation would be like? We can follow Hume's example and remain undecided agnostics or skeptics. Or, like Einstein and those who pray to be truly wise, we can continue in our pursuit of wisdom. That is, in more recent terms, in our pursuit of a grand unifying theory that gets everything right.

The Wonderful Myth Called Science

Perhaps a good way to ask where our 'whole intellectual world of ideas' comes from is to go back and ask again, "What or who is creating the colors we see as we read?" Is it possible that the entire stream of consciousness has a single cause? Those who are trying without success to attribute everything to their brains think so.

Why Einstein didn't give up.

In April 1950, *Scientific American* asked Einstein to write an account of his most recent 'extension' of general relativity theory. In what was his second last article written for the general public, Einstein spoke more of what drove him to continue working into his seventies than about the results of that work. In order to focus attention on his motives, he drew attention to what we might call 'the bulging library phenomenon.' Why are there more thousands of books and articles than ever being published by more writers than ever? Harry Bracken, for instance, began his 2002 addition to the literature on Descartes, the philosopher philosophers love to hate, by referring to the fact that an old bibliography of works published between 1800 and 1960 listed 3612 works on Descartes in its 510 pages. Forty years later, he was adding his own. (An excellent one.) Why?

Einstein phrased the question this way: "What, then, impels us to devise theory after theory? Why do we devise theories at all?" First, he wrote, we come upon new facts, and we enjoy trying to see how they reduce to old theories. But we also try to unify and simplify, comprehensively.

> ... There is another, more subtle motive of no less importance. This is the striving toward unification and simplification of the premises of the theory as a whole (i.e., Mach's principle of economy, interpreted as a logical principle).

> There exists a passion for comprehension, just as there exists a passion for music. That passion is rather common in children, but gets lost in most people later on. Without this passion, there would be neither mathematics nor natural

275

science. (A. Einstein, "On the Generalized Theory of Gravitation," *Scientific American*, April, 1950, p.13)

The purpose of this final chapter is to direct a spotlight of attention onto the question, "Does color exist?" It is perhaps the easiest of all the very difficult, unusual decisions forced on us by the recent discoveries we tend to call scientific. The reason for not breaking this final chapter into three or more chapters is to underscore the necessity for logical consistency in the way we understand all three parts of "I see colors."

Nothing less than striving for thorough-going consistency is consistent with the open-mindedness we associate with the wonderful myth called science-in-general.

E. REFLECTION MATERIAL

Abstract theory versus really knowing.

Einstein maintained that 'science is a refinement of everyday thinking.' Yet he also called naïve realism 'a plebeian illusion.' It logically follows that what we call science is built on a plebeian illusion. It is an illusion which is extraordinarily difficult to overcome, even for those eager to do so. The preceding pages offer abstract theory. But they can be used as virtual-reality therapy for virtual-reality illusion. We must use our naïve-realism to learn enough 'science' to overcome it! It calls for a huge paradigm-shift. Do you see these words? They're not words.

The rest of this chapter is not to argue. It is to assist in making this huge paradigm-shift.

How one person lost his ability to 'see.'

There are four everyday-thinking meanings of "see." The most important is the one we have in mind when we take the witness stand during a criminal trial and are asked to tell what we saw. It is the first meaning of "see" that a child learns. It's what we do with our eyes, not our nose or toes. Later on, we discover that we can, with eyes closed, picture what we saw a moment before, and that experience affords a

The Wonderful Myth Called Science

second meaning of "see," one that can also be conveyed by "visualize." (If what we saw a moment before was 'the sun,' what we experience when we close our eyes is so vivid that it has received a special name. It's called an after-image.)The third is the one used when we say "I can see that those first two meanings are not the same." It is synonymous with "understand." The fourth is the hybrid that combines seeing and thinking, i.e., "see" meaning "theory-ladenly perceive."

Our everyday knowledge about seeing is always on tap. Ask the question, "How do you know other people see the same colors you see?" Every non-blind freshman understands what that means without blinking an eye. But it is the wrong question! "How do I know colors exist?" is the right one: I can't see them if they don't exist. As for the colors others see, I'll never have apodictic evidence. It will always be a matter of faith. Of course, the very existence of other people will also be a matter of faith.

From the naïve-realist point of view, Aristotle's is the best analysis of what-I-sense. We see color, hear sound, smell odor, etc. On the basis of those, we can infer the sizes, shapes, distances, rest-or-motion, of bodies. As for the underlying reality, we must hypothesize. Guess.

When I first began creating a model for my naïve-realist worldview, that is, when I began studying St. Thomas and making explicit for myself what I had long believed implicitly, it never occurred to me that I could not see the colors of things. It never occurred to me that I saw light rather than colors.

Slowly, however, I became theory-blind to colors as such. I learned enough 'modern science' to incorporate into my thinking the formula, "I see the sky as blue because the atmosphere scatters the higher frequencies of sunlight which then travel to my eyes and cause me to see the sky as if it is that color." I could still see colors as well as ever. But my Big-Picture model no longer had room for them. It had room for light-absorbing and light-reflecting properties of material things such as the sky, clouds, grass, and so on. It had room for light, with its frequencies and wavelengths. It had room for eyes, retinas, afferent nerves, brain, and mind. But when I thought theoretically about color, my thoughts instantly switched to the items just referred to. Color as such, i.e., color as distinct from any and all of those items, had vanished.

'Sight' regained.

At the age of twenty–nine, without warning, I recovered. I suddenly realized that what I had naively mis-taken for the sun, clouds, and blue sky were effects in me, caused — I thought at the time — by my unseen brain, which was stimulated by impulses from eyes that 'transduce' light into nerve impulses. I suddenly realized that I was wrong when, as I was reading, I recited the formula, "The page reflects light to my eyes, causing the sensation in me that I (mis)interpret as color." I realized that the right way to think was expressed by "I am seeing colors, i.e., effects in me, and wrongly thinking of them as a nearby 'page' made of light-absorbing and light-reflecting lattices of colorless atoms." (To replace the illusion that I see them through my eyes, I often switch to, "I am directly aware of the colors in my mind." Nothing in my entire life prepared me for my 'recovery.'

The stunning realization that, like my 'hand,' 'the page' I was experiencing was the final effect in a cause-effect chain was initially terrifying. It meant that the links in the causal chain were out on the unseen side of this dazzling, beautiful, Technicolor vision. It also meant that the rest of the universe was on the unseen outside as well. But now I was ready to appreciate Berkeley as well as all the other thinkers who have tried to carry on from where Descartes left off.

A crucial corollary: some non-material things are extended.

One extremely important feature of my 'recovery' was this. In the thinking of St. Thomas and Descartes, nonmaterial things are non-extended. The mind is like a dimensionless point. Thoughts, too, are devoid of spatial dimensions. That changed when I discovered that what I thought I was seeing, viz., distant stars, a sun far off at the horizon, my two hands several inches from my face, were parts of a total visual field or TVF 'here in me.'

The Wonderful Myth Called Science

Whatever else may be the case, it is clear that immaterial sense-data and memory-images of them are extended. And so, this inner 'world' feels very roomy. Lots of space 'in here.' The visual sense-datum — the total visual field or TVF of spread-out colors — can only be described as being 'in front of me,' even though there is no distance between any part of it and me. It is clearly two-dimensional, and parts of it seem clearly three-dimensional. The TVF's parts can be inspected; they are not intro-spected. The sounds of birds and jackhammers, which every ounce of evidence proves are not out there, still seem to spread out around me, at times to be sharply located, at times moving from left to right or right to left or far to near or near to far.

I now describe the discovery with two reversible-direction metaphors. I've brought the spacious outer 'world' into my mind. Or I've expanded my mind to embrace this spacious inner 'world.' Either way, the mystery of mind only deepens.

Opposing 'phenomenological' descriptions.

To describe the radical shift involved in merging abstract theory and moment-by-moment experience, I often contrast certain features of my pre-twenty–nine theory and the features in my present view.

Description #1. Till I was twenty–nine, I was a naïve realist. But I had become a critical naïve realist. I understood Aristotle's theory of the senses. But I had used modern discoveries to update it.

Here is/was my update. Picture yourself watching a sunrise. The sun appears golden red, the clouds are edged with a similar color, and the sky is already grayish blue. Those 'colors' are from the light. Clouds and sky absorb some of the light and reflect or deflect the rest. The reflected light, invisible, reaches my eyes where the lenses focus it into upside-down patterns. The light causes a chemical alteration of the cones, nerve impulses are sent to the back part of my brain, and that explains my sensations. I see the physical things out in front of me, not invisible nerve impulses in the rear, occipital part of my brain. Depending on the frequencies of light reaching my eyes, I see the sun, clouds, and sky as colored, even though they are not colored as such. I see the sun as golden-red, the tinged clouds as mostly white, and the sky as blue, but, let me repeat, we now 'know' that color, so far as it refers to physical things, really means light-absorbing . . . etc.

What happens to what-I-see when I close my eyes? Not a thing happens to what-I-see. That is, nothing happens to the sun, clouds, or sky. Whatever happens, happens to me. I just stop seeing them. The effects Russell says are in me just cease. I cut off the light, that shuts off the nerve impulses, I stop being affected by them. I stop seeing the outside, unaffected, 'colored' things.

The scholastic theory, that the 'form' outside and inside are 'one,' seemed to remove any apparent contradiction. That solution — which I was still trying to understand better — was good enough for St. Thomas. Who was I to say it wasn't good enough?

Description #2. (Continue using the naïve-realist view of the physical world as a point of comparison.) What I see is a total visual field of patterned colors or colored patterns. The boundaries between the colors are sharp while I have my glasses on but blur slightly when I remove them. Part of the TVF I mis-take for the sun, other parts I mis-take for clouds and sky. Like my hand, the actual sun, clouds, and sky are on the far side of my TVF. They are invisible, they send invisible light to my unseen eyes which send unseen nerve impulses to my unseen brain, at which point this expanse of colors, shaped just as they seem to be on the movie or TV screen, is created in my mind.

But — and this is crucial — the visual field is not somewhere in the back part of my brain. To describe what I am visually aware of, I do not look behind me! I do not look inward or introspect to 'see' it. What I 'see' does not seem to be 'in me' at all. The only way to describe it is to say it seems to be out in front of me. I know it is flush up against me, just as truly as the sounds permeate my 'head' when I put stereophonic headphones on. If this makes it more difficult to fathom what the I I call "I" is, so be it. It's no easier to believe what physicists say about electrons orbiting the nuclei of atoms! But I believe both.

The Wonderful Myth Called Science

Colors as such are distinct from all of the other existents I believe in. What happens to the colors, what happens to what-I-see, what happens to the total visual field of patterned colors, when I close my eyes? What I see is annihilated. As for whatever is up in the sky, it is wholly unaffected when I close my eyes here below.

Now that I have accepted the theory of subatoms as the only reasonable way to think of physical things behind what I sense, it has become easier to reinterpret, as pragmatic-fiction laws-of-nature, what Schlieden & Schwann discovered about cells, what Crick & Watson discovered about DNA, what Ramon y Cajal discovered about neurons and Brown discovered about molecules and Rutherford, Millikan, Thomson, Compton, and Einstein discovered about photo-electricity. Physical things are dancing subatoms, not eyes, afferent nerves, or brains. As for the real source of the colors I experience, well, that calls for a decision about what "cause" should be taken to mean.

Reflecting on Helen Keller and the truly blind.

Every year we hear reports of new technologies to improve the lives of the deaf, quadriplegics, and other handicapped persons. Who has not heard of cochlear implants? Who is not astonished at reports of paralyzed patients learning to move a cursor on a monitor by creating brain-waves, thereby communicating by 'closed circuit captioning'? Currently, researchers are experimenting with tiny retinal implants to restore some vision to the blind.

What would happen if someone born blind, who had never experienced colors of any kind, were suddenly able to see? Helen Keller was nineteen months old when she was suddenly plunged into a world of total darkness and unbroken silence. If we ignore the disputes about whether or not she retained some memory-images of color and sound, we can carry out a thought-experiment. Suppose surgery could have restored her sight. What would she have experienced the first time the bandages were removed and she opened her eyes and could see what we see?

The answer helps us discover what Berkeley discovered before us. What Helen would have experienced is exactly what each of us experiences when we feel 'eye-opening' muscle sensations: a dazzling, beautiful, Technicolor panorama of color. The darkness of her world would have suddenly been flooded with 'light,' illuminating it the way a pitch dark room is suddenly 'lit up' when we turn on the lights. Close your eyes. Cover them to block out all the light. Wait till your eyes adjust to the darkness. Think about what you are experiencing. That is what 'being plunged into blindness' is like. Now open them. Think hard about what happened. And realize that what you experienced happened entirely in your mind.

No need to introspect. Only to notice. We do not see only a single star, the moon, a hand, or a page of print. Anything we pick out will be just part of an entire scene, a total visual field of colors, a nearly 180^0 panavision motion-picture. The whole field, if we close one eye and press the other, can be made to jiggle. Any part of the visual field, such as a 'page,' can be altered in a variety of ways with a magnifying glass. Or doubled with a mirror. Or turned upside down and made harder to read. (Try it.) The whole field turns upside down vis-à-vis our 'felt body' when we stand on our head or lean down to look through our 'felt legs.' The whole of it can be instantly annihilated by 'closing our eyes,' which closing will be signaled by felt-eye-muscle tightening. Do it. Open and close your eyes. And notice!*

But 'reading' the rows of tiny black figures is what seems (falsely) to cause a train of thought to go through our mind. That train of thought miraculously allows us to learn about the many ways the problem of

----

* Anyone able to study a television picture will see that it is a perfect substitute for a movie. No cars, only 'cars.' No people, only 'people.' Whereas cars and people don't change size, 'cars' and 'people' do as they 'come toward us,' 'go away,' 'disappear,' etc

perception has been 'solved' already. That reading suggests ways to eliminate the solutions that are impossible and to decide what's left must be the truth, however unbelievable it seems at first. And will seem unbelievable later on, too. Only seem, that is.

Helen Keller's 'higher education' via words, words, words.

"There is no frigate like a book." Or like newspapers. Without ever moving from our secluded ivory tower, said to be the favorite work-place for 'philosophers' but not 'scientists,' we can travel, not only around our cosmically tiny globe, but to the past of ancient history, to outer galaxies, through the terrestrial political and business worlds, to the latest science fiction about space-time, multiverses, and cats caught in the smaller-than-nano-second instant between life zand death. We can travel where we like on the magic carpet of words, words, words.

So, too, could the blind and deaf Helen Keller. When she suddenly lost her sight and hearing, her learning through heard sounds came to a screeching halt. In time, her learning resumed its normal pace in a way that was essentially like ours, only different. Reflection on the differences and similarities can powerfully aid in understanding our own higher learning. The difference is that which distinguishes tactile-somatic sensations from sounds. The similarity is that tactile sensations became words for her the way sounds and ciphers are for us. Besides announcing future sensations (the buzz or bite of a mosquito foretells an itch), they can be the occasion of increased knowledge of realities that transcend private sensations. A graphic account of her experience is found in *The Story of My Life.*

> I cannot recall what happened during the first months after my illness. I only know that I sat in my mother's lap or clung to her dress as she went about her household duties. My hands felt every object and observed every motion, and in this way I learned to know many things. Soon I felt the need of some communication with others and began to make crude signs. A shake of the head meant "No" and a nod, "Yes," a pull meant "Come" and a push, "Go." Was it bread that I wanted? Then I

283

would imitate the acts of cutting the slices and buttering them. If I wanted my mother to make ice-cream for dinner I made the sign for working the freezer and shivered, indicating cold. My mother, moreover, succeeded in making me understand a good deal. I always knew when she wished me to bring her something, and I would run upstairs or anywhere else she indicated. Indeed, I owe to her loving wisdom all that was bright and good in my long night. . .

The most important day I remember in all my life is the one on which my teacher, Anne Mansfield Sullivan, came to me. I am filled with wonder when I consider the immeasurable contrasts between the two lives which it connects. It was the third of March, 1887, three months before I was seven years old. . . .

The morning after my teacher came she led me into her room and gave me a doll. The little blind children at the Perkins Institute had sent it and Laura Bridgman had dressed it; but I did not know this until afterward. When I had played with it a little while, Miss Sullivan slowly spelled into my hand the word "d-o-l-l." I was at once interested in this finger play and tried to imitate it. When I finally succeeded in making the letters correctly I was flushed with childish pleasure and pride. Running downstairs to my mother I held up my hand and made the letters for doll. I did not know that I was spelling a word or even that words existed; I was simply making my fingers go in monkey-like imitation. In the days that followed I learned to spell in this uncomprehending way a great many words, among them pin, hat, cup and a few verbs like sit, stand and walk. But my teacher had been with me several weeks before I understood that everything has a name. . . .

We walked down the path to the well-house, attracted by the fragrance of the honeysuckle with which it was covered. Some one was drawing water and my teacher placed my hand under the spout. As the cool stream gushed over one hand she

The Wonderful Myth Called Science

spelled into the other the word water; first slowly, then rapidly. I stood still, my whole attention fixed upon the motions of her fingers. Suddenly I felt a misty consciousness as of something forgotten—a thrill of returning thought; and somehow the mystery of language was revealed to me. I knew then that "w-a-t-e-r" meant the wonderful cool something that was flowing over my hand. That living word awakened my soul, gave it light, hope, joy, set it free! There were barriers still, it is true, but barriers that could in time be swept away.

I left the well-house eager to learn. Everything had a name and each name gave birth to a new thought. (H. Keller, *The Story of My Life*, pp.14, 20-21, *passim*)

Helen Keller went on to graduate from Radcliffe College. Besides the tactile language Annie Sullivan used with her, Helen learned to feel-read braille, to feel-read lips-&-mouths. Besides English, she could feel-read works in French and German. She traveled around the world. After learning to speak words she could not hear she gave speeches in many countries. She wrote books and essays. Like Einstein's, her knowledge of the world expanded to such a degree that her 'writings' — how shall we analyze the full meaning of that! — can still serve as inspiration to us who sight-read them.

But a few lines from a 1958 *Readers Digest* abridgment of another book of hers are particularly important parallels to the quintalist thesis about the complex relations between thoughts, our stable imaged 'inner world,' the association-attachment of its items to auditory-image 'names,' and the final association-attachment — to the auditory-image 'names' — of visual-image 'print.' Our 'inner speech' consists of the auditory 'names.' Helen's consisted of tactile, not auditory, images.

Teacher wouldn't let the world about me be silent. I "heard" in my fingers the neigh of Prince, the saddle horse, the mooing of cows, the squeal of baby pigs. She brought me into sensory contact with everything that could be reached or felt—sunlight, the quivering of soap bubbles, the rustling of silk, the fury of a storm, the noises of insects, the creaking of a door, the voice of a loved one. To this day I cannot "command the uses of my soul" or stir my mind to action without the memory of Teacher's fingers upon my palm. (H. Keller, *Teacher*. chapter 5, abridged; *Readers Digest*, April, 1956.)

Once more: reading.

Till, like Helen Keller, we have become super-conscious of our own inner 'word-images,' it is highly unlikely that we will notice, as David Hume did, that we also have an inner, God's-eye view of a whole world of ideas.

> One would think the whole intellectual world of ideas was at once subjected to our view, and that we did nothing but pick out such as were most proper for our purpose. There may not, however, be any present, beside those very ideas, that are thus collected by a kind of magical faculty in the soul, which, though it be always most perfect in the greatest geniuses, and is properly what we call genius, is however inexplicable by the utmost efforts of human understanding. (*op. cit.*, p.31)

Inexplicable?

If, many years ago, this piece of paper had been held before your infant eyes, you would have seen exactly what you see now: hundreds of small figures against a white background. You might have stared at it for a moment. It might not have held your attention even that long. In either case, you would have comprehended nothing, not even that this was a page of writing you could not understand. Now, though you see nothing more than you saw then, what you see sets off chain reactions which occur so instantly that they are undetectable except by patient analysis.

You can also close your eyes and, as if by magic, visit 'the fields and vast palaces of memory' and relive long ago experiences with good friends. You can wander through countless worlds of imagination, ranging from Snow White and those Dwarfs to the most recent movie you have seen or novel you have read. You can recall historical eras you

286

have read about, zoom out to the edges of infinite space, then zoom back in to the incredible events taking place within the atom.

That inner world, parts of which were paraded onto the stage of your explicit consciousness just now because you saw certain tiny black marks, is a world within your own mind. Is there no connection, no bridge, between that inner world and the outer world of objective reality? Earlier in life, you'd reply "If I wish, I can put down this book and experience the real world by looking, listening, smelling, tasting, and touching." The thousands of tiny black figures your eyes have scanned on the pages of this book have brought before you the reasoning which says to you, "That reply is naïve."

A new paradigm: thoughts as proposals

When David Hume had his 'Magic Moment,' he felt that "the whole intellectual world of ideas" was spread out before him. As for how that whole world of ideas got into his mind, David's answer was, "They are copies of what I have experienced."

But what about Helen Keller? Her writings show that she too had a whole intellectual world of ideas at her command. How did she get hers? They were not copies of what she heard, which is the reason she put quotes around "heard." Had her ideas about this world and the vast cosmos beyond come from what she sensed, they would have been copies of tactile sensations, beginning with the 'feel' of Annie Sullivan's fingers in her palm. Try to imagine an 'inner model' of the world composed of tactile sense-data!

'Nature' is full of tricks. She make us believe that we get ideas of stars, trees, squirrels, and other things by sensing them. But we don't. And so, Einstein proposed his provisional we-create-our-ideas model for them.

But what model can we create for the ideas we seem to get from other people, but don't? William James's thesis, namely, that our 'higher' knowledge comes — seemingly — via words, words, words, suggests another model. When we hear others speak words or read the words they write, thoughts come to us, and it often seems that persons are revealing new things to us. If we apply traditional logic to those thoughts, we may describe them as propositions. Inasmuch as we are free to assent to them, dissent from them, or withhold any decision about them, we can rename them "proposals." Propositions or proposals that seemingly come from others' words are 'things being told to us.'

But who is doing the telling or proposing?

Unfinished business

Aristotle's 'natural philosophy' represented his attempt to explain change. His physics, cosmology, biology, and psychology were his particular-science efforts to explain the things discussed in this non-text 'text.' Quintalism reduces a vast array of changes to the behavior of material subatoms and to humans' streams of immaterial conscious experience.

But, just as Aristotle recognized at the end of his physics that changes do not explain themselves, the argument here should bring pursuers of a complete unified theory to the realization that the predictably unpredictable dances of subatoms and the understandable flow of conscious experience have not been given a causal explanation in this text. Recall Wittgenstein's verdict: "The whole modern conception of the world is founded on the illusion that the so-called laws of nature are the explanations of natural phenomena."

What that means, of course, is that those in love with the wonderful myth, that a thing most call "Science" offers causal explanations for the dances of subatoms and the flow of conscious experience, have further to go in their quest for wisdom.

One requirement for any complete unified theory should now be obvious. Any theory that pretends to explain the learning which Einstein, Bohr, Darwin, and James did, must explain David Hume's 'magic moment and Helen Keller's learning, and whatever thoughts or things have just been told to you.

EPILOGUE

Descartes & Dreaming

Let this book end with a reflection for anyone who doesn't know how to be absolutely certain 'this' is not all a dream.

No one should ever worry about losing the ability to tell the difference between being awake and knowing what's going on, and being asleep and dreaming about what's not going on. You, for instance, have known the difference from the time you were six. In fact, out of the mouths of babes . . . One young child, asked whether she was certain she was not just dreaming, replied: "I might be day-dreaming, but I'm not night-dreaming."

But if you do begin to worry, go to see *Bambi* the next time the movie comes to a theater near you. Suspend your disbelief and immerse yourself in the story. Fall asleep. When you wake, try to notice the obvious change as you pass from one state to the other. Then see if you can recall whether or not you got engrossed in a dream while you were asleep. If so, weigh the evidence provided by the series of experiences to see whether or not the following is true.

First, the characters and events in the movie-story existed only in your imagination. The story was 'evoked' by the real colors and sounds which you really experienced. Those real colors and sounds, unlike the characters and events of the fictitious story produced by your imagination, trace back to the wizardry of modern motion-picture technology. Whenever you wish, those identical colors and sounds can be re-produced in exactly the same sequence by the same technology that produced them the first time. Your sense experience will always fit that

Epilogue

hypothesis, even though that idea of what produces your sense experience is false.

Secondly, the dream experience is radically different. None of the events in your dream involved real, flesh-and-blood people. But also, none of the events can be re-produced at will in exactly the same sequence. And you'll never even have an experience of what, if anything, you guess was the cause of the dream experience. If you guess it was your unconscious, it would be silly to insist you were consciously conscious of your unconscious. If it was your brain, that too is something you'll never, ever be conscious of. (Close your eyes and see if you can tell which part of your brain is producing your thoughts right now.)

Thirdly, once you have reasoned about the difference between being on-the-edge-of-your-seat-awake during a movie and being-asleep-dreaming during a movie, ask yourself what you would reply to a friend who leaned over during the movie and whispered: "Do you really believe those are animals you see up on the screen, or that they think and talk just like us?" (Incidentally, Bentham asked the wrong question. It isn't whether animals can feel pain. It's whether they exist.)

Finish by memorizing the above directions. Use them to decisively reject any future worry about whether you are awake and conscious or this is all a dream.

There is a bonus. You'll also never have to doubt whether or not you are awake while you are re-reading this book. (You are awake, no?)

Of course, converting your personal opinions into personal scientific opinions is not a matter of seeing. It's knowing. That includes knowing what you're seeing. And remembering what you have seen. Remembering is thinking when the thinking is of the past. Descartes paid more attention to thinking than to seeing, and that was good. But thinking correctly about seeing is a huge part of knowing. Hume's analysis of cause-effect thinking showed that thinking correctly about seeing is mostly remembering.

The Wonderful Myth Called Science

Warning. This book is not for one-time reading. It is an entire course. To prepare for a comprehensive exam at the end of a course, all of the material must be reviewed. It cannot be reviewed at the beginning, not until the last day of class finishes. Or compare this book to a novel. Early chapters of a novel often cannot be fully understood without its final pages. For instance, what sense can a reader make of Lord Marchmain's gesture at end of *Brideshead Revisited* who reads only that page or who does not know anything of Waugh's mind? What sense can a reader make of the first two thirds of Waugh's novel who has no access to the last third? So it is here. Unless you see how the chapters' main topics and the central ideas in the quotes from Einstein all fit together, you'll never get into Einstein's mind. You'll not get into his mind if you do not get into Kant's mind, which cannot be done without getting into Descartes' mind, which cannot possibly be done by anyone who does not notice the complexity of the everyday, common-sense thinking which alone has made it possible for you to read this book.

Index

INDEX

INDEX

INDEX

INDEX

OTHER TITLES FROM SOLAS *Press*

Distributor: Baker & Taylor Tele. 908 541 7508. Fax 908 704 9315

People from the Dawn: Religion, Homeland, and Privacy in Australian Aboriginal Culture

By W. E. H. Stanner and John Hilary Martin

ISBN 9781893426986   Publication Date: October 2001

Paperback   168 pages   Price $24.95

The human equality of aboriginal populations was established in the 1500's when the Conquistadores were the dominant force in America. However, centuries later, the full personhood of the indigenous Australians was questioned. Here is an Australian work that shows the unexpected depth and sophistication of Australian Aboriginal culture

The Reality of Myth

By John Hilary Martin

ISBN 9781893426993. Publication Date: September 2001

Paperback 138 pages Price $22.95

Reductionism and skepticism are uneasy bedfellows of myth! Yet myth is necessary for human creativity. Intuiting, imagining, and rationalizing utilize myth. But beyond utility, a clear and distinct idea of myth's contribution to human knowledge is needed. Martin makes this connection simply by grounding mythic symbolism through analogy. Using well-known mythic stories he shows that far from being an escape from reality, myth is an escape into a fuller reality.

The exposition is woven around the 'creation' myths in American Indian culture, the Bible and the 'Dreaming' of Australian aborigines. Here is a seminal work that will open our eyes to the essential role of myth in culture and religion; its unsuspected role in science; and creativity.

The True Church: The Path which Led a Protestant Lawyer to the Catholic Church.

By Peter H. Burnett

ISBN 9781893426740 Publication Date October 2004

Casebound 768 pages Price $37.95

Given pioneering California's stress on the practical arts it will come as a surprise that, its first Governor was a scholar. Peter Burnett was a magnificent pioneer of the "old west." Besides being a lawyer, soldier, newspaper editor, farmer, Supreme Court judge, businessman, politician, and statesman he was a scholar. In this book Burnett applies plain juristical logic to the question of the true church. His contemporary Orestes Brownson, said of Burnett, "Through him California has made a more glorious contribution to the Union than all the gold of her mines..." Cardinal Dulles says of this work, "Burnett, applying the principles of Anglo-American law to Scripture and the Fathers of the Church, produced a remarkably full and impressive apologia...." Cardinal Levada says "His (Burnett's) work is a tour de force."

This book is relevant today, as well as having profound historical value. The issues are indeed current –interpretation of Scripture, the role of reason in religious faith, the need for a tangible church, achieving certainty, and so on.

VISIT OUR WEB SITE:

www.solaspress.com

ORDER DIRECTLY:

SOLAS Press, P.O. Box 4066, Antioch CA 90509 USA

Toll Free 1 888 407 SOLAS Tele 1 925 978 9781

Fax 1 925 978 2599 E-mail info@solaspress.com

www.ingramcontent.com/pod-product-compliance
Lightning Source LLC
Chambersburg PA
CBHW021030210326
41598CB00016B/968